AF595472

Advanced Information Processing Methods and Their Applications

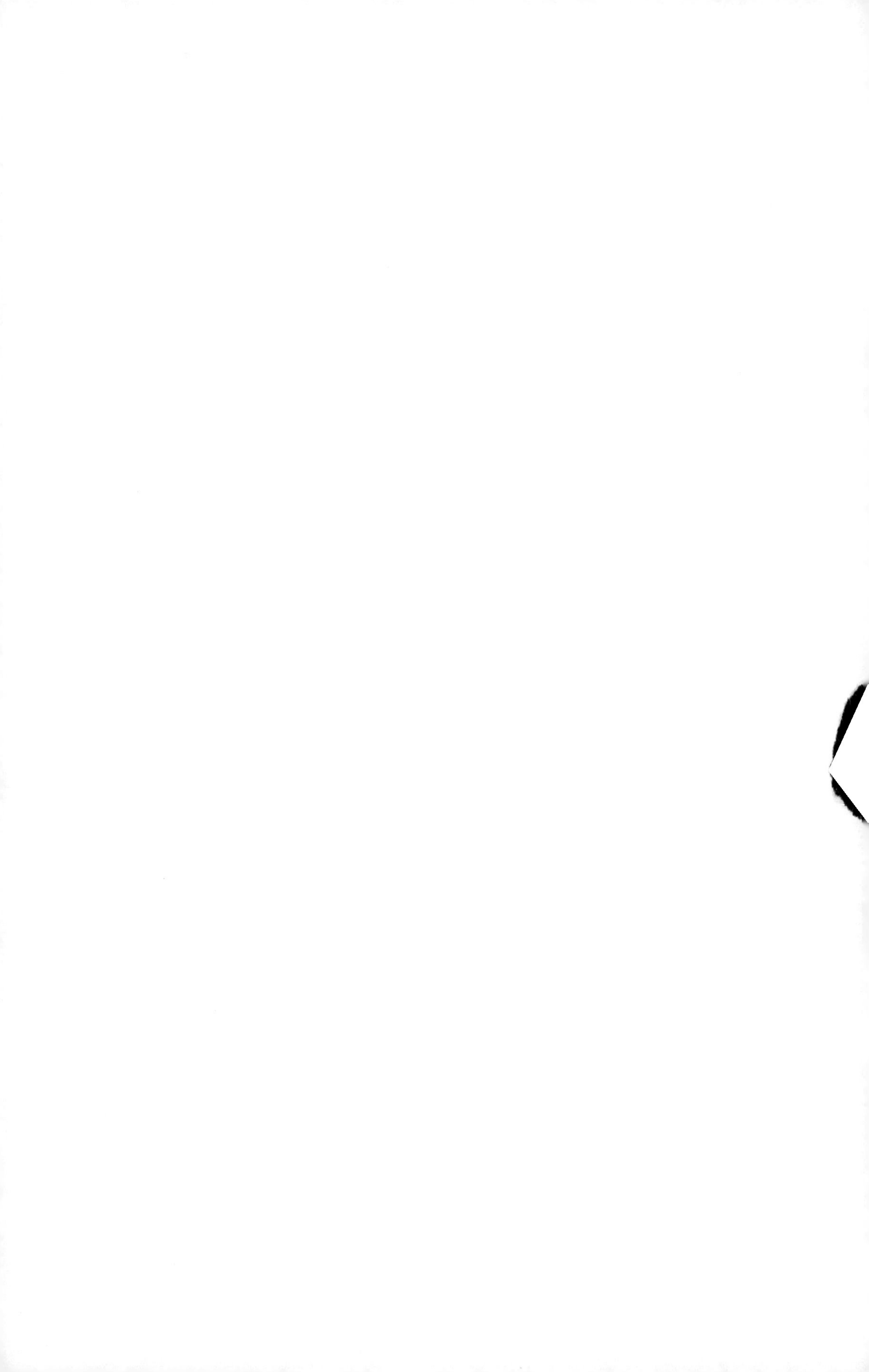

Advanced Information Processing Methods and Their Applications

Editor

Pavel Lyakhov

MDPI • Basel • Beijing • Wuhan • Barcelona • Belgrade • Manchester • Tokyo • Cluj • Tianjin

Editor
Pavel Lyakhov
North-Caucasus Federal University
Russia

Editorial Office
MDPI
St. Alban-Anlage 66
4052 Basel, Switzerland

This is a reprint of articles from the Special Issue published online in the open access journal *Applied Sciences* (ISSN 2076-3417) (available at: https://www.mdpi.com/journal/applsci/special_issues/information_processing_methods).

For citation purposes, cite each article independently as indicated on the article page online and as indicated below:

LastName, A.A.; LastName, B.B.; LastName, C.C. Article Title. *Journal Name* **Year**, *Volume Number*, Page Range.

ISBN 978-3-0365-5515-7 (Hbk)
ISBN 978-3-0365-5516-4 (PDF)

Contents

About the Editor

Pavel Lyakhov

Prof. Dr. Pavel Lyakhov, PhD in Computer Sciences, Head of the Department of Mathematical Modeling, Faculty of Mathematics and Information Technology named after Professor N.I. Chervyakov, North-Caucasus Federal University.

Pavel Lyakhov is currently the head of the following scientific projects: "High-performance devices for digital processing of medical images based on parallel mathematics" and "Hardware accelerators for digital processing of three-dimensional medical images using scaled filters and parallel modular computing". His interests include high performance computing, residual class system arithmetic, digital signal processing, digital image processing, machine learning and artificial intelligence, medical imaging and custom hardware development. Pavel Lyakhov is the author of more than 150 published scientific papers and teaching materials, including 55 articles in publications indexed in Scopus and Web of Science, 18 copyright certificates and patents, and 2 textbooks.

Preface to "Advanced Information Processing Methods and Their Applications"

The rapid development of information technology opens up new opportunities for quality improvement in many areas of human activity. Modernity is characterized by a significant increase in the volume of extracted and processed information, which leads to the problem of developing new approaches to organizing computations, including neurocomputing and quantum computing. Digital circuits are also under active development, especially in improving performance and reducing power consumption for use in mobile and embedded devices. New problem-oriented solutions based on FPGA and ASIC are constantly being developed for a variety of applications. New digital signal, image, and video processing devices must meet the growing practical needs for high speed and high quality of work. The widespread use of machine learning methods and new methods for big data processing will help humans in many areas. One of the most promising areas for the application of modern IT technologies is biomedical data processing. An interesting issue in modern science is the development of new brain–computer interfaces. Another important application of computer science is medicine, especially in the context of the development of new tools for the diagnosis and support of patients with COVID-19.

The latest technological developments in the areas listed above will be shared through this Special Issue. We invite researchers and investigators to contribute their original research or review articles to this Special Issue.

Pavel Lyakhov
Editor

Editorial

Special Issue on Advanced Information Processing Methods and Their Applications

Pavel Lyakhov [1,2]

[1] Department of Mathematical Modeling, North-Caucasus Federal University, 355017 Stavropol, Russia; ljahov@mail.ru
[2] Department of Modular Computing and Artificial Intelligence, North-Caucasus Center for Mathematical Research, 355017 Stavropol, Russia

The rapid development of information technology opens up new opportunities in many areas of human activity. Modernity is characterized by a significant increase in the volume of extracted and processed information, which leads to the problem of developing new approaches to the organization of calculations, including neurocomputing and quantum computing. Digital circuits are also under active development, especially in terms of improving performance and reducing power consumption for use in mobile and embedded devices. New problem-oriented hardware solutions are constantly being developed for a wide variety of applications. Another important application of information technology is in medicine, especially in the context of developing new tools for diagnosing and supporting patients with COVID.

This Special Issue has collected and presented breakthrough research on information processing methods and their applications. Particular attention is paid to the study of the mathematical foundations of information processing methods, quantum computing, artificial intelligence, digital image processing, and the use of information technologies in medicine.

A total of five research papers in various fields of information processing methods and their applications are presented in this Special Issue. Lyakhov et al. [1] introduced a new approach for the determination of atrial fibrillation on electrocardiogram signals based on neural networks with wavelet-based preprocessing. Cariow et al. [2] reported an algorithm for the fast multiplication of Kaluza numbers which are used in various fields of data processing, including digital signal and image processing, machine graphics, telecommunications, and cryptography. Vahabi et al. [3] developed a new design and implementation of novel efficient full adder-subtractor circuits based on quantum-dot cellular automata technology. Minahil et al. [4] proposed patch-wise infrared and visible image fusion using spatial adaptive weights. Isabona et al. [5] developed a multilayer perceptron neural network for optimal predictive modeling in urban microcellular radio environments.

Although submissions for this Special Issue have been closed, more in-depth research in the information processing methods and their applications continues to address the challenges we face today, such as the development of neural network and quantum approaches, search for new effective software and hardware solutions, implementation of advanced methods of information processing in practice.

Citation: Lyakhov, P. Special Issue on Advanced Information Processing Methods and Their Applications. *Appl. Sci.* **2022**, *12*, 9090. https://doi.org/10.3390/app12189090

Received: 7 September 2022
Accepted: 8 September 2022
Published: 9 September 2022

Funding: This research received no external funding.

Acknowledgments: Thanks to all the authors and peer reviewers for their valuable contributions to this Special Issue 'Advanced Information Processing Methods and Their Applications'. I would also like to express my gratitude to all the staff and people involved in this Special Issue.

Conflicts of Interest: The author declare no conflict of interest.

References

1. Lyakhov, P.; Kiladze, M.; Lyakhova, U. System for Neural Network Determination of Atrial Fibrillation on ECG Signals with Wavelet-Based Preprocessing. *Appl. Sci.* **2021**, *11*, 7213. [CrossRef]
2. Cariow, A.; Cariowa, G.; Paplinski, J.P. An Algorithm for Fast Multiplication of Kaluza Numbers. *Appl. Sci.* **2021**, *11*, 8203. [CrossRef]
3. Vahabi, M.; Lyakhov, P.; Bahar, A.N. Design and Implementation of Novel Efficient Full Adder/Subtractor Circuits Based on Quantum-Dot Cellular Automata Technology. *Appl. Sci.* **2021**, *11*, 8717. [CrossRef]
4. Minahil, S.; Kim, J.-H.; Hwang, Y. Patch-Wise Infrared and Visible Image Fusion Using Spatial Adaptive Weights. *Appl. Sci.* **2021**, *11*, 9255. [CrossRef]
5. Isabona, J.; Imoize, A.L.; Ojo, S.; Karunwi, O.; Kim, Y.; Lee, C.-C.; Li, C.-T. Development of a Multilayer Perceptron Neural Network for Optimal Predictive Modeling in Urban Microcellular Radio Environments. *Appl. Sci.* **2022**, *12*, 5713. [CrossRef]

MDPI

Article

Development of a Multilayer Perceptron Neural Network for Optimal Predictive Modeling in Urban Microcellular Radio Environments

Joseph Isabona [1], Agbotiname Lucky Imoize [2,3], Stephen Ojo [4], Olukayode Karunwi [5], Yongsung Kim [6,*], Cheng-Chi Lee [7,8,*] and Chun-Ta Li [9]

1 Department of Physics, Federal University Lokoja, Lokoja 260101, Nigeria; joseph.isabona@fulokoja.edu.ng
2 Department of Electrical and Electronics Engineering, Faculty of Engineering, University of Lagos, Akoka, Lagos 100213, Nigeria; aimoize@unilag.edu.ng
3 Department of Electrical Engineering and Information Technology, Institute of Digital Communication, Ruhr University, 44801 Bochum, Germany
4 Department of Electrical and Computer Engineering, College of Engineering, Anderson University, Anderson, SC 29621, USA; sojo@andersonuniversity.edu
5 College of Arts and Sciences, Anderson University, Anderson, SC 29621, USA; okarunwi@andersonuniversity.edu
6 Department of Technology Education, Chungnam National University, Daejeon 34134, Korea
7 Research and Development Center for Physical Education, Health and Information Technology, Department of Library and Information Science, Fu Jen Catholic University, New Taipei City 24205, Taiwan
8 Department of Computer Science and Information Engineering, Asia University, Taichung City 41354, Taiwan
9 Department of Information Management, Tainan University of Technology, 529 Zhongzheng Road, Tainan City 710, Taiwan; th0040@mail.tut.edu.tw
* Correspondence: kys1001@cnu.ac.kr (Y.K.); cclee@mail.fju.edu.tw (C.-C.L.)

Citation: Isabona, J.; Imoize, A.L.; Ojo, S.; Karunwi, O.; Kim, Y.; Lee, C.-C.; Li, C.-T. Development of a Multilayer Perceptron Neural Network for Optimal Predictive Modeling in Urban Microcellular Radio Environments. *Appl. Sci.* **2022**, *12*, 5713. https://doi.org/10.3390/app12115713

Academic Editors: Pavel Lyakhov and Amalia Miliou

Received: 12 April 2022
Accepted: 2 June 2022
Published: 3 June 2022

Abstract: Modern cellular communication networks are already being perturbed by large and steadily increasing mobile subscribers in high demand for better service quality. To constantly and reliably deploy and optimally manage such mobile cellular networks, the radio signal attenuation loss between the path lengths of a base transmitter and the mobile station receiver must be appropriately estimated. Although many log-distance-based linear models for path loss prediction in wireless cellular networks exist, radio frequency planning requires advanced non-linear models for more accurate predictive path loss estimation, particularly for complex microcellular environments. The precision of the conventional models on path loss prediction has been reported in several works, generally ranging from 8–12 dB in terms of Root Mean Square Error (RMSE), which is too high compared to the acceptable error limit between 0 and 6 dB. Toward this end, the need for near-precise machine learning-based path loss prediction models becomes imperative. This work develops a distinctive multi-layer perception (MLP) neural network-based path loss model with well-structured implementation network architecture, empowered with the grid search-based hyperparameter tuning method. The proposed model is designed for optimal path loss approximation between mobile station and base station. The hyperparameters examined include the neuron number, learning rate and hidden layers number. In detail, the developed MLP model prediction accuracy level using different learning and training algorithms with the tuned best values of the hyperparameters have been applied for extensive path loss experimental datasets. The experimental path loss data is acquired via a field drive test conducted over an operational 4G LTE network in an urban microcellular environment. The results were assessed using several first-order statistical performance indicators. The results show that prediction errors of the proposed MLP model compared favourably with measured data and were better than those obtained using conventional log-distance-based path loss models.

Keywords: path loss models; log-distance models; neural networks models; MLP-based models; optimal predictive modelling; multi-layer perception neural network; urban microcellular radio networks

1. Introduction

Path loss models are unique prediction models employed by telecom network engineers to estimate the signal coverage area being served by a given transmitter during networking and management [1–3]. However, developing these signal path loss models with the optimal accuracy it deserves is a complex and significant problem in the planning of telecommunication networks. The conventional log-distance-based statistical models available in the literature, such as the cluster factor model, COST 234 Hata Model, free space model, Hata model and Lee models, lack accuracy for realistic path loss prediction applications in cellular mobile networks environments [4–11]. The aforementioned fundamental limitation of the conventional models is usually very pronounced when the respective models have been applied in cellular radio environments other than the developed and designed environment [12,13]. This scenario is mainly due to dissimilarities and variations in environmental formations (hilly, mountainous or quasi-plain), weather conditions, soil electrical properties and terrain type (open, rural, suburban or urban) that exist in different radio propagation locations, cities and countries [14–23]. For example, the Hata loss model was developed based on extensive practical measurements carried out in Japan at transmission frequency ranges of 150 to 1920 MHz and macrocellular communication distances of 1 to 100 km, with the mobile station and base station antenna heights of 2 to 3 and 30 to 1000 m, respectively [24]. This model, including other aforementioned conventional ones, is also generally limited in capturing the non-linear relationship between the independent variable (e.g., path loss) and dependent variable (e.g., distance) [25]. The precision of these conventional models on path loss prediction has been reported in many previous works to generally range from 8–12 dB in terms of Root Mean Square Error (RMSE), which is exceptionally higher than the acceptable values [9,14,15].

Recently, an Artificial Neural Network (ANN), a unique artificial intelligence soft computing and modelling technique, has been acknowledged and proven to solve function approximation and pattern classification problems [26]. Some ANN models exist in the literature; the key ones are Radial Basis Function models, Multilayer Perception Models, Generalized Neural Network models, etc. Among these models, the MLP ANN models have stood out most recently because they are very robust and popular for learning, function approximation, and pattern classification [27–30]. The MLP ANN possess many robust algorithms that can be explored to carry out more proficient adaptive nonlinear statistical modelling over the classical logistic regression methods [14–23] that are frequently engaged in developing predictive models. This robustness can be ascribed to their acknowledged special ability to learn, predict, and classify non-linear data using experience and preceding samples introduced to the network model. Huang [31] also noted that the MLP is characteristically good for input-output data mapping. Generally, a clear-cut underlining capability of ANN-based models over the conventional log-distance-based models is their large degrees of freedom structure which provides means for fitting many datasets with non-linear or linear correlation patterns. The concept of intelligent-based ANN models for optimal and adaptive prognostic estimations of path losses was introduced to surmount the limitations of existing empirically and deterministically developed log-distance models [32,33]. In the paper, the availability of manifold resourceful training algorithms and hypermeters of MLP ANN that can be tuned to further boost its extrapolative data analysis is worth exploring in this paper for optimal predictive modelling. Hence, the "Development of a Multilayer Perception Neural Network for Optimal Predictive Modeling in Urban Microcellular Radio Environments is self-evident." Other key robust advantages of the general ANNs are highlighted in Section 2.3.

Though several ANN models exist in the literature, a critical, challenging task remains in developing and using them appropriately through the correct selection of its network structural design with the required input elements and hyperparameters to solve peculiar predictive mapping and functional problems. The quest to address this issue is the leading motivation for this research paper. However, one primary challenge in using the MLP model is correctly selecting its network architecture with the required input elements (hyperparameters) to solve a particular mapping problem [34]. Another critical challenge with neural network models is the problem of determining the input data variables that must correlate with the target variables [35,36].

This paper develops a distinctive MLP-based path loss model with well-structured implementation network architecture, empowered with the grid search-based hyperparameter tuning method for optimal path loss approximation between mobile-station and base-station path lengths. The hyperparameters include the neuron number, learning rate and hidden layers number. In detail, the developed MLP model prediction accuracy level using different learning and training algorithms with the tuned best values of the hyperparameters have been applied for extensive path loss experimental datasets. The datasets were acquired via field drive tests conducted in Long Term Evolution (LTE) in urban microcellular radio networks. For the development and implementation of the MLP-ANNs model, we utilized version 2018a of the MATLAB neural networks toolbox. The toolbox provides the required user interface, algorithm and platform to train, test, validate, visualize and simulate networks with the desired number of layers, neurons and activation functions.

In particular, the contributions of this paper are summarized as follows:

- A distinctive MLP neural network-based path loss model with well-structured implementation network architecture, embedded with the precise transfer function, neuron number, learning algorithm and hidden layers, is developed for optimal path loss approximation between mobile-station and base-station path lengths.
- The developed MLP neural network model was tested and validated for realistic path loss prediction using extensive experimental signal attenuation loss datasets acquired via field drive test conducted over LTE networks in urban microcellular radio environments and tested using first-order statistical performance indicators.
- Optimization of the projected model via hyperparameter tuning leveraging the grid search algorithm analyses of the experimental path loss data.
- Optimal prediction efficacy of the developed MLP neural network model compared with well-structured implementation architecture over standard log-distance models using several first-order statistical performance indicators.

The remainder of this work is structured in the following manner. Section 2 outlines the background information, such as radio propagation mechanisms, log-distance-based path loss prediction models, and the basis of artificial neural networks (ANN). Section 3 presents the methodology detailing the neural network implementation for predictive modelling. Section 4 provides the results, analysis, evaluation of the introduced neural network model at different study locations, comparison of the developed neural network model with log-distance models, and discussions. Finally, the conclusion is given in Section 5.

2. Theoretical Background

The theoretical background covers the radio propagation mechanism, log-distance-based path loss prediction models and artificial neural network systems.

2.1. Radio Propagation Mechanism

When radio signals travel, which are a form of electromagnetic waves, they interact with the media and objects they travel through. In the sequence of their interaction, the radio signals become weaker owing to refraction, reflection, diffraction, absorption and other propagation phenomena. The resultant effect of all the phenomena on propagated signals is signal propagation loss. The characteristics of the pathway or medium through which the radio signals travel determine the amount of propagation loss and the quality of the received signal that is attainable at the receiving terminal. Radio propagation loss is also governed by other sundry elements, particularly the transmitter power, receiver sensitivity and general antenna parameters such as antenna gain, antenna height and receiver location [1,2,37,38].

The prominent factors that influence the number of signal path losses in a medium include diffraction, reflection, refraction, scattering and absorption, to mention a few. For example, diffraction arises when radio waves collide with huge obstacles compared to the propagating signal wavelengths. Moreover, diffraction occurs when radio signals bend around objects, especially those with sharp edges. This alteration often empowers the received radio signal energy to spread around the boundaries of the obstructing object [39,40]. Diffraction is also influenced by the phase, amplitude, pathway and frequency of the transmitted waves.

The environment in which the radio frequency signals travel (or are propagated) will undoubtedly negatively impact the signal. For example, radio wave signals and propagation loss vary extensively in correspondence to the terrain landscape, building structures and population density. Marshy, damp and sandy terrain also attenuate radio signals, primarily propagated low-frequency signals. In other words, signals travel faster over conducive terrain than in sandy and marshy or damp terrains.

2.2. Log-Distance-Based Path Loss Prediction Models

Generally, path loss models are a set of mathematical models, expressions, resources and algorithms used for signal attenuation loss prediction between the paths of a base transmitter and the mobile station receiver. These models are helpful planning tools that assist the radio network designers of cellular telecommunication systems in optimally positioning base station transmitters to meet the desired signal coverage level and service quality requirements of the networks.

The predictive performance of any path loss model is determined by the resultant prediction accuracy with actual field measured loss data.

The log-distance-based path loss models are models whose average power loss logarithmically depends on distance (transmission path length) intertwined with a propagation exponent modelling parameter. The propagation exponent is usually employed to account for a specific radio propagation environment. They can also be described as simplified models that attempt to model variations, fluctuations and attenuations in the received signal power. Examples of log-distance-based models include the Walficsh–Ikegami, Walficsh Bertoni, cluster factor, COST 234 Hata, Hata Okumura, SUI model, Lee model, Egli Model and others [41]. Though these models have varying frequency validity thresholds, different correction factors have been applied to ease their applicability at the tested frequency band. Detailed descriptions of these models are contained in [14,17,42].

2.3. Artificial Neural Networks (ANNs)

ANNs, also popularly referred to as artificial neural systems, are efficient computing systems or relatively simple computational models founded on the neural organization of the brain with functional changing parameters to process information effectively. ANNs are distinctive and robust non-linear statistical data modeling networks wherein reasonably simple connections between inputs and outputs nodes alignments are established. According to Robert Hecht-Nielsen, the first inventor of the neurocomputer, a neural network can be defined as *"a computing system made up of several simple, highly interconnected processing elements, which process information by their dynamic state response external inputs"*. The processing elements are called neurons. The neuron is the special mathematical function that captures and organizes information according to the neural network architecture.

Some of the essential features or advantages of ANNs are [31,34–36].

a. High-speed computation adeptness
b. Global interconnection of network processing elements
c. Robust distributed and asynchronous parallel processing
d. High adaptability to non-linear input-data mapping
e. Robust noise tolerance
f. High fault tolerance utilizing Redundant Information Coding
g. Robust in providing real-time operations

h. High potential for hardware application
i. Capable of deriving meaning from the imprecise or complicated dataset
j. High capacity to learn, recall and generalize input data training pattern

3. Methodology

As mentioned earlier, the ANN model possesses many robust training algorithms and hyperparameters that can be explored to conduct proficient adaptive nonlinear statistical modelling over the classical logistic regression methods. This section contains the materials and method explored to develop the proposed MLPANN-based model with well-structured implementation network architecture, empowered with the right hyperparameter tuning algorithm for optimal predictive analysis of practical path loss data. The stepwise exploratory method explored to develop the proposed MLPANN-based model is highlighted as follows:

a. Acquire the path loss datasets
b. Preprocessing the dataset and splitting
c. Obtain the MLP neural network model.
d. Identify the adaptive learning algorithms for MLP neural network model training and testing
e. Identify the modelling hyperparameters
f. Select a hyperparameter tuning algorithm for the model (e.g., Bayesian optimization search, grid search, etc.).
g. Obtain a set of the tuned best hyperparameter values.
h. Train the model to obtain the best hyperparameter values combination.
i. Appraise and validate the accuracy of the training process.
j. Repeat the process to optimal configuration and best-desired results for the model.
k. Engage the model with well-structured implementation network architecture, learning algorithm and set of the tuned best hyperparameter values to conductive predictive path loss modelling.

3.1. Data Collection

The field measurement was conducted to acquire live signal data around three Long Term Evolution (LTE) transceiver base station antennas for one year (i.e., 12 months). The measurement took one year to cater to the study locations' seasonal variations and three LTE transceiver base station antennas operating at 2600 MHz with 10 MHz bandwidth [43]. The transceiver base station antennas (called NodeBs) are sectorized with 17.5 dBi gain and 43 dBm transmit power. The LTE network belongs to one of the major GSM/WCDMA/HSPA/LTE telecom service providers operating across major towns, villages and cities in Nigeria. The measurements were performed with field test tools with TEMS application tools for radio spectrum analysis. The test tools, some of which are displayed in Figure 1, include a Rover car, scanner, two Samsung mobile phones, and an HP lap, were explored to assess the performance of eNode B over the LTE radio air interface by connecting mobile phones directly to the Node B transmitters. To obtain the eNode B locations and delineate measurement data locations/information, the Global Positioning System (GPS) equipment was employed. The path loss data to be predicted are related to the acquired radio signal data by the measured path loss data where PL_{mea}(dB), values have been obtained from the measured signal, RSRP (dBm) by Equation (1):

$$PL_{mea}(\text{dB}) = EIRP + G_A - RSRP_{meas} \tag{1}$$

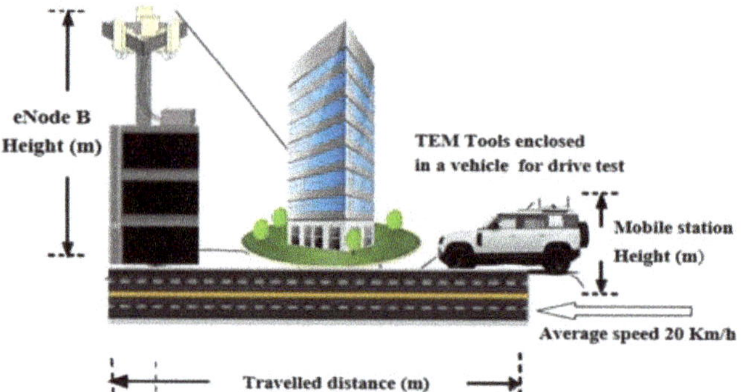

Figure 1. Illustration of the TEMS Drive Test Measurement System.

With *EIRP* calculated as Equation (2):

$$EIRP = P_{TX} + G_{TX} - CL_{TX} \tag{2}$$

where G_{TX} and G_A are the base station (BS) transmit antenna gain and receiver (MS) antenna gain, respectively, P_{TX} is the transmitted power, and CL_{TX}, denotes transmission cable loss, all in dB. Table 1 reveals some of the key BS antenna site parameters acquired during the field drive test for calculation.

Table 1. Measurement Path Loss Computation Parameters.

Parameters	Site 1	Site 2	Site 3
BS Operation Transmitting Frequency (MHz)	2600	2600	2600
BS Antenna Height (m)	28	30	45
MS Antenna Height (m)	1.5	1.5	1.5
BS antenna gain (dBi)	17.5	17.5	17.5
MS antenna gain (dBi)	0	0	0
MS Transmit power (dB)	30	30	30
BS Transmit power (dB)	43	43	43
Transmitter cable loss (dB)	0.5	0.5	0.5
Feeder Loss (dB)	3	3	3

3.2. The MLP Neural Network Model

The first step towards effectively engaging neural networks for predictive modelling is to know the exact type you want to use and determine network architecture. This paper considers the most robust and special type of neural network: the multi-layer perceptrons (MLP). A single perceptron (LP) has limitations in terms of input-desired output mapping capability. This limitation is because it only contains a single neuron per adaptable synaptic weights and bias; thus, it is only proficient in catering to ridge-like function, notwithstanding the type activation function explored [42]. The above limitation can be catered to using an MLP neural network with more source nodes with data input and output layers sandwiched with hidden layers nodes. Multiple layers of neurons in the MLP network provide enhanced input-desired output mapping capability.

Figure 2 displays a structure of a classic feedforward MLP network model composed of $g_1, g_2, \ldots g_I$, inputs, and predicted output, $(y_1, y_2, \ldots, y_N)$ with k_h hidden nodes, h number. The respective weights connecting the input and hidden layer, as well as the weights connecting the hidden layer and the output layer, are designated by w_{ij}^1 and w_{jn}^2, while C_j indicates the hidden nodes thresholds. The network learns the correlation between input datasets and predicted output feedback by varying weight and bias values. Accordingly, the MLP network predicted output in correspondence to jth neurons with the kth node could be articulated as (3):

$$\hat{y}_n(t) = \sum_{j=1}^{k_h} w_j^2 F\left(\sum w_{ij}^1 g_i(t) + c_j\right) \tag{3}$$

for $1 \leq n \prec m, 1 \leq j \prec k_h, (w_j, j = 0, 1, \ldots, k_h), (w_{ij}, i = 0, 1, \ldots, m; j = 0, 1, \ldots, k_h)$

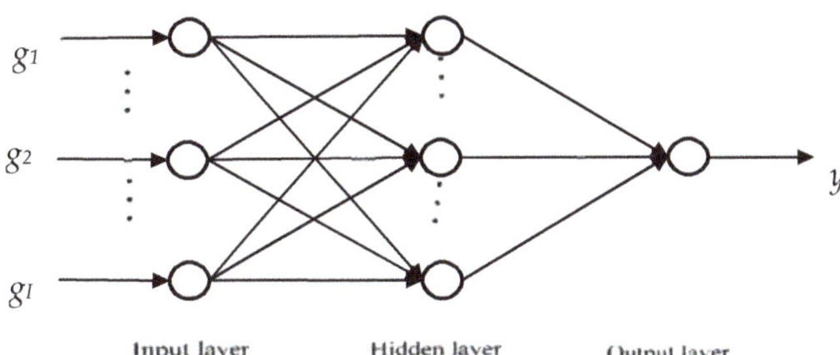

Figure 2. Scheme of a three-layered ANN multi-layer perceptron.

where:

m, h and k_h indicate the input node number, hidden node and hidden node number, respectively; i designate input to j hidden layer neuron.

The F $(\cdot)$ in Equation (1) denotes the sigmoid activation function, an import function usually utilized in the MLP network. It can be defined by Equation (4):

$$F(a) = \frac{1}{1 + e^{-a}} \tag{4}$$

where $F(x)$ is at all times in the range $[-1, 1]$, with $F(a)$ being a set of real numbers.

The weights w_{ij}^1 and w_{jn}^2, including the threshold C_j, are unknown and thus can be chosen to update and reduce the error during prediction. The prediction error can be expressed by employing the expression (5):

$$\varepsilon_n = \frac{1}{2} \sum_{n=1} (y_n - \hat{y}_n)^2 \tag{5}$$

where,

y_n and $\hat{y}_n$ represent the target (i.e., actual) data and their predicted output; and $n = 1$ $\ldots$, N, with N indicating the actual data sample number.

In MLP training, the error verve for assessing the network learning improvement related to convergence speed is the generalized aggregate error values. It is often computed using mean square error (MSE). The MSE can be obtained from the least square formation of Equation (6).

$$MSE = \frac{1}{N} \sum_{n=1}^{N} (y_n - \hat{y}_n)^2 \tag{6}$$

In this work, the feedforward MLP network model explored for path loss predictive modelling is displayed in Figure 3.

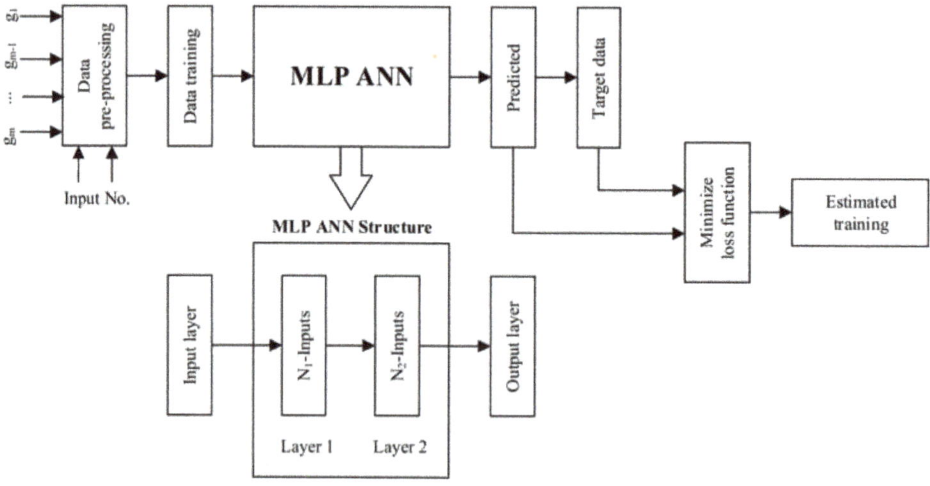

Figure 3. ANN MLP Model Structure for Path Loss Prediction.

3.3. *MLP Modelling Parameters and Search Space*

Hyperparameters are a special set of regulating parameters that the NN model utilizes for the adaptive learning process in data training and testing. The special parameters may be categorical, continuous or integer variables whose values range are usually lower and upper bounded. Thus, there exists several MLPs directly impacting the predictive modelling. They include the hidden layers number, neurons number in the hidden layer, transfer function, etc. A summarised description of the transfer function is given in the following subsection.

3.3.1. The Hidden Layers

Deciding the number of the hidden layers is one of the most important issues while investigating the neural network architecture for predictive modelling and data mining. Using too many hidden layers can result in poor generalization and complex neural network training. According to authors in [44–47], two hidden layers, combined with m output neurons, are adequate for a neural network to learn N data samples and produce negligibly minor errors.

Previous studies have examined the suitability of several machine models for path loss predictions as contained in [48–50]. The need to overcome the problems of empirical models when used for path loss predictions led to an artificial neural network [49]. ANN path loss prediction models were also more efficient and easier to deploy than deterministic models [51]. In [52,53], analyses of empirical models with different propagation features were performed, and the model with the lowest RMSE value was then compared with the prediction from ANN. The ANN-based path loss prediction model produced a much lower value of MSE upon validation. In [54], a multi-layer perceptron neural network was introduced for path loss prediction. The MLP network was then trained with a backpropagation algorithm. The MLP-based prediction was compared with predictions from analytical models, and the results indicated the former to be efficient for radio network planning and optimization.

ANN was also used for path loss prediction in urban areas [55]. The work explored the effect of the various input parameters and the environmental terrains on the robustness of the path loss prediction. One key finding from the study is that the accuracy of the signal prediction model increases with more input parameters: the greater the number of features, the greater the system's accuracy. This trend is because machine learning algorithms thrive well with the availability of large datasets. The model is trained with the help of the data and, in this case, the input features. An ANN-based path loss model at 800 MHz and 1800 MHz introduced in [47,48] were input for longitude, latitude, distance, elevation, clutter height and elevation. The ANN method in [56] outperforms the Support Vector Machine (SVM)- and the Random Forest (RF)-based predictions.

In [57], an artificial neural network was used for path loss prediction in a smart campus environment at 1800 MHz. There were two hidden layers for this network, and the performance of the network outperforms the prediction made by using RF. Moreover, in [58–60], several machine learning-based prediction models were introduced for signal predictions for wireless sensor networks. The machine learning-based prediction model in [61,62] also produced the lowest values of RMSE when compared to the other analytical models in a wireless sensor network.

3.3.2. Neurons Number in the Hidden Layers

Determining the neurons number in the hidden layers remains an integral part of the inclusive neural network architecture. An inadequate neurons number in the hidden layers can lead to an underfitting problem. Underfitting arises once there are insufficient neurons number in the hidden layers to learn or detect the signals satisfactorily, especially in a multifaceted dataset.

On the other hand, using too many neurons can lead to overfitting problems. Overfitting occurs once the neural network contains too much information processing capacity problem. It can also result in excessive time increase during neural network training. The amount of training time can increase to the point that it is impossible to train the neural network adequately. It is evident that some give and take must be grasped between too few and too many neurons number in the hidden layers.

3.3.3. Transfer Function

The transfer function is a singular, monotonically increasing and differentiable function used for translating the input data signals to produce the final output signals of a neuron. The transfer function is fundamental to the concrete concept of neural networks mainly for two key reasons. First, without activation functions, the entire organization of the neural network will be similar to a typical linear function that cannot learn non-linear relationships. Second, transfer function styles and graces the main computation accomplished by neural networks.

3.3.4. Learning Rate

The learning rate *is* another vital hyperparameter that regulates or fine tunes the weights of NN in relation to the loss gradient. Its value must be cautiously chosen to support both optimization and generalization robustly. A too-large learning rate value can cause the entire learning process to jump over minima. Similarly, a too-small learning rate value can make the entire learning process too long to converge, resulting in it being trapped in negative and spurious local minima.

3.4. Hyperparameter Tuning

Hyperparameter tuning or optimization expresses the robust procedure of identifying and finding the best feasible values of hyperparameters for a machine learning model to attain the desired resultant modeling outcome. Popular hyperparameter tuning algorithms in the literature include random search, grid search and Bayesian optimization search. In this paper, the last two methods are considered. Grid search is a standard hyperparameter optimization technique wherein a list of critical parameters is selected with attached feasible values for each parameter, followed by training the model for every single blend and then choosing the values that yield the most desired resultant outcome. The Bayesian optimization method is a special sequential model-based optimization (also known as Bayesian optimization), utilizing the 'Bayes Theorem' to conduct an automatic hyperparameter search. Particularly, the Bayesian optimization search algorithm utilizes the upshot from the preceding iteration to select the next hyperparameter values.

3.5. MLP Learning Algorithms

The training algorithm used for a neural network system to learn and solve a problem is essential. The correct training algorithm from the available sundry types depends on diverse factors, including data sample size, task type, training time constraints, precision/accuracy requirements, etc. It is demanding to find out which training algorithm will produce the most satisfactory results. For example, suppose it is a predictive modelling task with function approximation. In that case, the dominant or most common ones are backpropagation training algorithms, which involve carrying out computations backwards over the network to fine-tune the weights and minimize performance error.

3.6. MLP Network Model Implementation Process

MATLAB is a distinctive programming language with a multi-exemplar numerical computing environment and a user interface. It provides easy matrix manipulation, graphical multi-domain simulation, figurative computing, creative functions plotting, excellent data mining, easy algorithms implementation, etc. MATLAB allows access to optional toolbox uses. The neural network toolbox has special tools for model-based design, implementation, visualization and simulation of neural networks. MATLAB is employed to encode the script files for the MLP network model predictive training, testing and quantitative evaluation in this work. The program code for conventional path loss calculation and assessment is also explored. The proposed MLP neural model consists of five input nodes and one output node. Flowchart for executing proposed predictive modelling with ANN while training and testing are shown in Figure 4.

As mentioned earlier, for practical and optimal application of MLP network for predictive modelling purposes, the right choice of the learning algorithm, and selection of the network processing elements such as the number of neurons, number of network layers and transfer functions, are crucial. For example, a network with insufficient neurons number in a hidden layer may fail to capture complex links between target output and input variables. Conversely, if the number of neurons allotted in the network hidden layer is too many, the network would likely follow the latent noise in the dataset owing to over-parameterization, and this, in turn, can lead to awkward generalization and poor predictive modelling of the original data [63,64]. Therefore, the determination of the hidden layers number and their number of neurons is performed by trial and random selection. However, for conciseness and the need to attain optimal neural network training and testing, the search for the required number of neurons and hidden layers in the network layers were narrowed down to 2–50 and 1–3, respectively.

Generally, if a particular algorithm performs well during the dataset training but flops in the aspect of generalization, we refer to it as overfitting. To improve generalization (or prevent overfitting) during the path loss data training with each of the NN algorithms, we employed input/target transformations and early stopping techniques. Thus, the inputs and targets datasets were scaled to reside in the range $[-1, 1]$ to enhance training and testing speed. Moreover, the early stopping measures were engaged for training and testing to avoid overtraining, eliminate contemptuous impact stimulated by the initial values, and develop robust adaptive predictive ability. Although many learning algorithms are available for MLP neural network training and testing in MATLAB software, it is demanding to identify which algorithm works best for a given predictive modelling problem concerning convergence speed and accuracy [16]. Therefore, an exhaustive search is employed in this study to accelerate the convergence and evaluate the impact of all the available learning algorithms during network training. The respective learning algorithms assessed to develop an optimal MLP network predictive model and their weight adaptation techniques are listed in Table 2. The target of the weight adaptation is to determine the optimum weight update for the input-output data array pair during training.

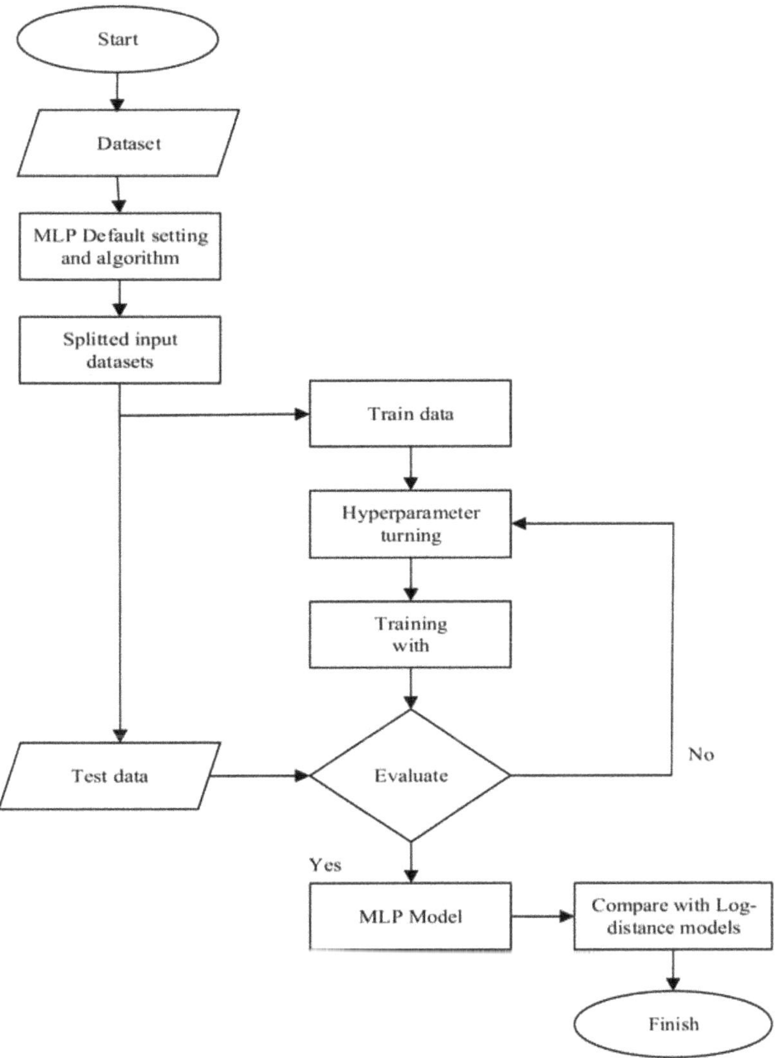

Figure 4. Flowchart for the execution of Proposed Predictive Modelling with MLP Neural Network.

Table 2. Respective MLP Network Learning Algorithm.

Learning Algorithm	Weight Adaptation	Training Acronym
Levenberg–Marquardt (lm)	The weight of the lm algorithm is updated via: $w_{q+1} = w_q - (H' + \eta I)^{-1} \nabla \varepsilon_q$ where the Hessian matrix, $H' = J^T J$.	trainlm
Bayesian Regularization (br)	The br approach involves the modification of the performance function, ε_q plus regularisation term $F(w) = \beta \varepsilon_q + \alpha \varepsilon_w$ where β & α are special function parameters and a regularisation term, $\varepsilon_w = \|w\|^2$	trainbr
Polak–Ribiere Conjugate Gradient (cgp)	The weight of cgp algorithm is updated by: $w_{q+1} = w_q + a_q p_q$ With $p_q = -g_q + b_q p_{q-1}$ and $b_q = \frac{\Delta g_{q-1}^T g_q}{g_{q-1}^T g_{q-1}}$	traincgp
Fletcher-Powell Conjugate Gradient (cgf)	The weight of scg algorithm is updated by: $w_{q+1} = w_q + a_q p_q$ With $p_q = -g_q + b_q p_{q-1}$ and $b_q = \frac{g_{q-1}^T g_q}{g_{q-1}^T g_{q-1}}$	traincgf
Scaled Conjugate Gradient (scg)	The weight of scg algorithm is updated by: $w_{q+1} = w_q + a_q p_q$	trainscg
Resilient Backpropagation (rp)	With RP algorithm, weight update is by $w_{q+1} = w_q - sign\left(\frac{\Delta \varepsilon_q}{\Delta w_q}\right).\Delta(q)$ where $\Delta(q)$ = individual step size for weight adaptation	trainrp
BFGS Quasi-Newton (bfg)	bfg weight update is accomplished via $w_{q+1} = w_q - H^{-1} g_q$ where H_q^{-1} indicates the Hessian matrix inversion on iteration q.	trainbfg
Conjugate gradient with Powell/Beale Restarts (cgb)	The cgb employs update search direction by: $g_{q-1} g_q = 0.2\|g_q\|^2$	traincgb
Gradient Descent with Adaptive Learning Rate (gda)	gda weight update is accomplish via $w_{q+1} = w_q + \frac{\Delta \varepsilon}{\eta_{q+1} \Delta w}_q$ where α indicates the user-defined momentum factor and it ranges from 0 to 1	taingda
Gradient Descent Variable Learning Rate (gdx)	With gdx algorithm, weight update is by $w_{q+1} = w_q \eta_{q+1} - sign\left(\frac{\Delta \varepsilon_q}{\Delta w_q}\right).\Delta(q)$ where $\Delta(q)$ = individual step size for weight adaptation and η_q = variable learning rate	traingdx
One Step Secant (oss)	With oss algorithm, weight update is by $w_{q+1} = w_q - H_q^{-1} g_q$ where H_q = Hessian matrix (2nd derivatives)	trainoss
Gradient Descent with Momentum (gdm)	gdm weight update is accomplish via $w_{q+1} - w_q - \eta g_q + \alpha \frac{\Delta \varepsilon}{\Delta w}_{q-1}$ where α indicates the user-defined momentum factor and it ranges from 0 to 1	traingdm
Gradient Descent (gd)	With gm algorithm, weight update is by $w_{q+1} = w_q - \eta g_q$ where η = learning rate	traingd

4. Results and Discussions

As mentioned earlier, many factors directly impact the development of an excellent back-propagation neural network predictive model, especially with a trial and error method, as adopted in this work. They include training algorithm, hidden layers number, neurons number in the hidden layer, transfer function, momentum term, learning rate, etc. Here, the concentration is on the training algorithm, hidden layers number and neurons number in the hidden layer. To obtain the predictive path loss modelling results, first we divided the path loss data into portions and employed the grid search-based hyperparameter tuning method to generate configurations for the parted data chunks for training and testing. The performance results of the proposed method were evaluated and reported for each machine learning algorithm described in Table 1, using Mean Absolute Error (MAE), Mean Square Error (MSE), Root Mean Square Error (RMSE), Coefficient of Efficiency (COE), Correlation Coefficient (R) and Standard Deviation Error (STD) [65].

Secondly, the predictive path loss modelling was conducted for three study locations using the standard Bayesian optimization for hyperparameter tuning. The results were compared to our first results using the grid search-based hyperparameter tuning method.

4.1. Neurons Number Impact

Determining the neurons number in the hidden layers also remains integral in the inclusive neural network architecture. An inadequate neurons number in the hidden layers can lead to an underfitting problem. Underfitting arises once there is insufficient neuron number in the hidden layers to satisfactorily learn or detect the signals, especially in a multi-faceted dataset. On the other hand, making use of too many neurons can lead to overfitting problems. Overfitting takes place once the neural network contains too much information processing. It can also result in excessive time increase during neural network training. The amount of training time can increase to the point that it is impossible to adequately train the neural network. It is very clear at this point that some give and take must be grasped between too few and too many neurons number in the hidden layers. Accordingly, by starting with the fastest training algorithm, which is Levenberg–Marquardt (lm), the network was trained and tested with 2, 5, 10, 15, 20, 25, 30, 35, 40, 45 and 50 neurons. This is to ascertain their incremental impact on its performance. Table 3, Figures 5 and 6 display the detailed overall predictive performance of each neurons number using different error statistics. As seen in Table 3, it was expected that a continuing increase in neurons number per layer would result in an upturn in the resolution of the neural network prediction fitting pattern to the measured loss data. At 30 neurons per layer, the neural network model had a good enough fit to the measured loss data with an MAE value of 2, RMSE value of 2.71, STD value of 1.82, R-value of 95 (%) and COE value of 90 (%). Increasing the neuron number to 50 showed no performance improvement, as seen in Figure 6.

Table 3. Neurons Number impact on MLP Network Predictive Modelling with LM.

	Training		Testing		Overall Performance				
No of Neurons	**MSE**	**R**	**MSE**	**R**	**MAE**	**STD**	**RMSE**	**R**	**COE (%)**
2	12.91	0.8874	23.22	0.7913	2.96	2.41	3.81	0.8637	75
5	12.22	0.8643	11.93	0.9205	2.92	2.40	3.78	0.8637	75
10	7.32	0.9320	8.01	0.8900	2.96	2.38	3.80	0.9251	86
15	7.95	0.9229	15.72	0.8339	2.41	1.92	3.08	0.9117	83
20	8.59	0.9247	13.79	0.9186	2.34	1.95	3.05	0.9148	84
25	7.83	0.9296	53.08	0.4865	2.47	2.96	3.85	0.8635	75
30	5.08	0.9591	17.06	0.8558	2.00	1.82	2.71	0.9499	90
35	4.88	0.9531	22.45	0.8482	1.95	2.06	2.84	0.9255	86
40	7.29	0.9360	14.31	0.7973	2.16	2.36	3.20	0.9132	84
45	7.95	0.9315	11.06	0.8257	2.23	2.07	3.04	0.9143	84
50	3.62	0.9673	21.07	0.7624	2.13	3.66	4.24	0.8498	72

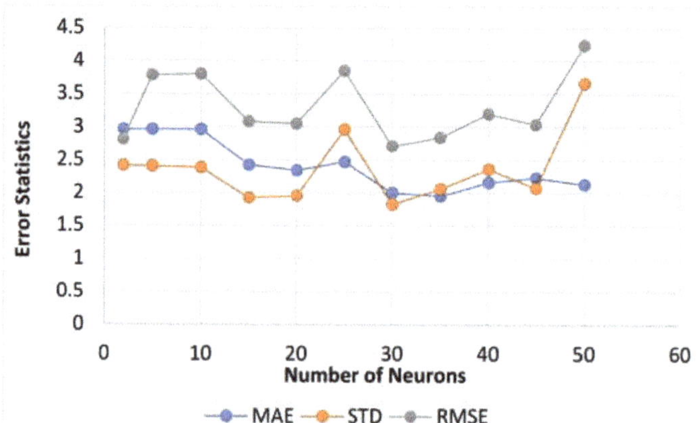

Figure 5. Overall Performance Error Statistics with MAE, STD and RMSE.

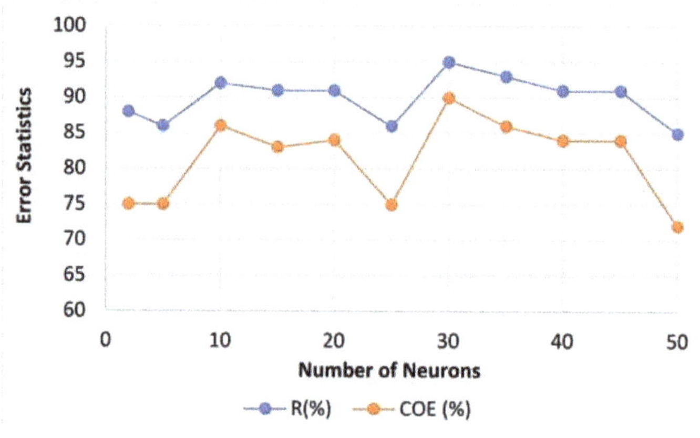

Figure 6. Overall Performance Error Statistics with R (%) and COE (%).

4.2. Transfer Function Impact

The transfer function is a singular monotonically increasing and differentiable function used for translating the input data signals to produce the final output signals of a neuron. The transfer function is fundamental to the concrete concept of neural networks mainly for three key reasons. Firstly, without activation functions, the entire organization of the neural network will be similar to an ordinary linear function that cannot learn non-linear relationships. Secondly, transfer function styles grace the main computation accomplished by neural networks. Thirdly, transfer functions possess the proclivity to boost the learning rate and formation patterns in datasets. Thus, the choice of the right transfer (activation) function also positively influences the performance of the NN training algorithm. Table 4 presents the list of sigmoid transfer functions employed in this study to ascertain the stability of the proposed neural network model. The performance of sigmoid transfer functions in terms of MAE, MSE, RMSE, R and STD are also displayed in Table 4. The standard deviation (STD) statistics with one, two and three layers of training are given in Figure 7, and the Root Mean Error performance statistics with one, two and three layers of training are shown in Figure 8.

Table 4. Transfer Functions used for network training/testing and their performance.

	Training		Testing		Overall Performance			
Transfer Function	**MSE**	**R**	**MSE**	**R**	**MAE**	**STD**	**RMSE**	**R**
purelin	19.74	0.8614	18.29	0.8692	3.36	2.65	4.28	0.8636
tansig	6.82	0.9366	26.07	0.7865	2.19	2.45	3.29	0.9073
logsig	4.62	0.9594	23.09	0.7812	2.15	2.30	3.15	0.9106
purelin-purelin	20.33	0.8715	13.09	0.8964	3.59	2.75	4.52	0.8637
purelin-tagsig	7.83	0.9325	12.20	0.8539	2.38	1.80	2.99	0.9173
purelin-logsig	9.18	0.9173	20.24	0.8755	2.52	2.02	3.24	0.9026
tansig-tansig	3.18	0.9697	20.17	0.8643	1.82	1.89	2.63	0.9365
tansig-logsig	3.23	0.9697	23.65	0.8094	1.82	2.18	2.84	0.9266
tansig-purelin	6.39	0.7622	10.28	0.6907	2.04	1.82	2.73	0.9315
logsig-logsig	4.82	0.9511	10.31	0.9382	1.76	1.63	2.40	0.9290
logsig-tansig	3.97	0.9669	12.96	0.8467	1.60	1.70	2.34	0.9499
logsig-pureline	6.01	0.9430	16.45	0.8471	2.18	1.91	2.90	0.9222
purelin-purelin-purelin	13.07	0.8804	18.34	0.8413	2.93	2.39	3.78	0.8637
tansig-tangsig-tansig	5.87	0.9487	8.89	0.9181	1.88	1.78	2.59	0.9283
tansig-tansig-purelin	6.38	0.9337	17.89	0.8607	2.17	2.06	2.76	0.9200
tangsig-tansig-logsig	5.11	0.9568	8.98	0.8907	1.92	1.93	2.72	0.9318
logsig-logsig-logsig	7.18	0.9358	13.86	0.9187	2.15	1.85	2.83	0.9285
logsig-logsig-purelin	9.12	0.9238	98.48	0.6384	2.91	4.03	4.97	0.8049
logsig-tansig-tansig	5.34	0.9489	7.52	0.9429	2.18	1.92	2.90	0.9245
logsig-tansig-logsig	6.90	0.9428	3.97	0.9467	1.98	1.75	2.67	0.9340
tansig-logsig-tansig	6.53	0.9433	3.68	0.9468	1.97	1.56	2.47	0.9445
logsic-tansig-purelin	6.92	0.9392	11.34	0.9093	2.09	2.02	2.90	0.9222
purelin-logsig-tansig	25.57	0.8583	25.94	0.8707	3.28	3.58	4.86	0.8597

Figure 7. Standard deviation (STD) statistics with one, two and three layers of training.

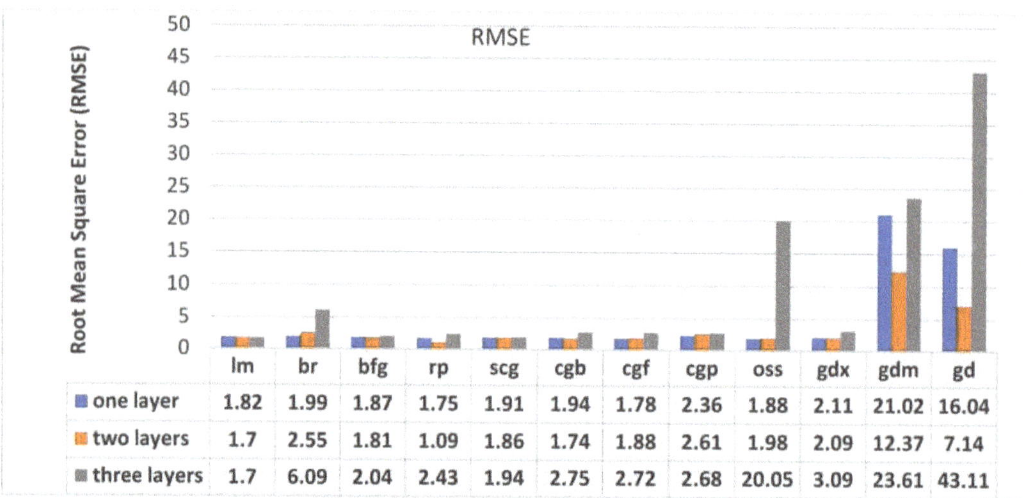

	lm	br	bfg	rp	scg	cgb	cgf	cgp	oss	gdx	gdm	gd
one layer	1.82	1.99	1.87	1.75	1.91	1.94	1.78	2.36	1.88	2.11	21.02	16.04
two layers	1.7	2.55	1.81	1.09	1.86	1.74	1.88	2.61	1.98	2.09	12.37	7.14
three layers	1.7	6.09	2.04	2.43	1.94	2.75	2.72	2.68	20.05	3.09	23.61	43.11

Figure 8. Root Mean Error performance statistics with one, two and three layers of training.

4.3. Training Algorithm and Hidden Layers Number Impact

The learning algorithm and hidden layers number also significantly impact the success of the neural network in coming up with a healthy predictive model. Therefore, deciding on the number of the hidden layer is one of the most important issues that come up while investigating the neural network architecture for predictive modelling and data mining. Using too many hidden layers can result in poor generalization and complex neural network training. According to the work [31], two hidden layers, combined with m output neurons, are adequate for a neural network to learn N data samples and produce negligibly small errors.

Here, the impact of many learning algorithms has been studied with one, two and three hidden layers numbers. Interestingly, results reveal that the two-layered network is superior to one-layered and three-layered layer network for all the 12 learning algorithms investigated. Interestingly, results show that the neural network architecture trained using lm (i.e., the Levenberg–Marquardt training algorithm) with two hidden-layer sizes and logsig-transit transfer function gave the best performance. Table 5 displays detailed network training/testing error statistics results and hidden layer numbers for each learning algorithm.

4.4. Performance of Grid Search Algorithm and Bayesian Optimisation Search Algorithm for Hyperparameter Tuning

The hyperparameter tuning process has a weighty influence on neural network learning performance. Given the computational resources requisite, the hyperparameters of high relevance receive superior usage in the hyperparameter tuning process. Hyperparameters with a more robust influence on weights are more effective during neural network training. Thus, the appropriate choice of hyperparameters selected for neural network model training influences the network training and performance.

As displayed in Table 6, the proposed MLP neural network model results with grid search algorithm-based hyperparameter tuning (optimization) are compared with those obtained using the traditional Bayesian optimization search-based hyperparameter tuning approach for path loss data predictive analysis using location one as a case study. We have reported the results attained for lm and br learning for brevity. While the grid search-based hyperparameter tuning performs an in-depth and comprehensive search on the hyperparameters in a stepwise manner as set specified by users with limited search space, the Bayesian search-based hyperparameter tuning performs a sequential-based search on the hyperparameters via several trials, without the user having preliminary information of the hyper-parameters distribution. From the results in Table 6, it is clear that the proposed MLP neural network model with grid search algorithm-based hyperparameter tuning outperforms the ones obtained using Bayesian Optimization search-based hyperparameter tuning. Next, Section 4.5 is to apply the proposed MLP neural network model with grid search algorithm-based hyperparameter tuning for detailed predictive path loss analysis across the three study locations.

Table 5. Learning Algorithm and Hidden Layers Number impact on MLP Network.

		Training		Testing		Overall Performance			
Training Algorithm	No of Hidden Layers	MSE	R	MSE	R	MAE	STD	MSE	R
lm	1	7.65	0.9359	11.55	0.9033	2.20	1.82	2.86	0.9248
	2	3.97	0.9669	12.96	0.8467	1.60	1.70	2.34	0.9499
	3	3.94	0.9632	9.31	0.9311	1.71	1.70	2.41	0.9473
Br	1	3.72	0.9660	26.59	0.7560	1.79	1.99	2.68	0.9340
	2	0.57	0.9953	47.07	0.7562	1.05	2.55	2.76	0.9408
	3	0.92	0.9919	25.68	0.3075	1.73	6.09	6.33	0.7557
Bfg	1	7.88	0.9252	14.88	0.9070	2.56	1.87	3.17	0.9076
	2	9.12	0.9092	12.92	0.8869	2.42	1.81	3.03	0.9149
	3	10.04	0.9159	10.93	0.8425	2.45	2.04	3.20	0.9045
Rp	1	7.68	0.9316	5.03	0.9601	2.09	1.75	2.73	0.9323
	2	4.38	0.9339	17.02	0.9123	1.90	1.09	2.60	0.9456
	3	5.03	0.9052	28.59	0.8816	2.69	2.43	3.63	0.8755
Scg	1	7.85	0.9282	11.08	0.8608	2.26	1.91	2.97	0.9185
	2	6.69	0.9373	13.52	0.7531	2.26	1.86	2.92	0.9209
	3	11.16	0.8951	9.65	0.8951	2.66	1.94	3.29	0.9005
Cgb	1	9.01	0.9168	14.75	0.8882	2.49	1.94	3.31	0.9089
	2	8.30	0.9293	8.89	0.9233	2.36	1.74	1.95	0.9202
	3	16.36	0.8460	29.72	0.8814	3.26	2.75	4.27	0.8510
Cgf	1	8.44	0.9186	12.65	0.9079	2.42	1.78	3.00	0.9165
	2	9.15	0.9152	8.93	0.9204	2.33	1.88	2.99	0.9168
	3	17.84	0.8274	21.14	0.8596	3.39	2.72	4.35	0.8449
Cgp	1	18.22	0.8496	23.49	0.7818	3.57	2.36	4.28	0.8426
	2	17.19	0.8472	11.26	0.9039	3.05	2.61	4.01	0.8482
	3	38.02	0.7463	36.79	0.7099	3.06	2.61	4.00	0.7486
Oss	1	9.53	0.9167	12.55	0.8703	2.70	1.88	3.28	0.9015
	2	9.06	0.9161	9.77	0.9116	2.49	1.98	2.49	0.9055
	3	10.16	0.9101	16.90	0.8084	2.59	20.05	3.30	0.8984
Gdx	1	11.82	0.8805	15.02	0.8584	2.91	2.11	3.59	0.8794
	2	9.59	0.9060	18.15	0.8430	2.54	2.09	3.29	0.8982
	3	18.60	0.8286	12.07	0.8252	3.50	3.09	4.66	0.8113
Gdm	1	1.42×10^3	0.8008	1.26×10^3	0.7041	30.26	21.02	36.84	0.7870
	2	447.60	0.7801	571.30	0.5385	16.97	12.37	21.01	0.8034
	3	1.22×10^3	0.4340	1.5×10^3	1.5×10^3	26.67	23.61	36.38	0.4504
Gd	1	2.94×10^3	0.7238	3.08×10^3	0.5882	53.81	16.04	56.15	0.6286
	2	655.14	0.8583	724.47	0.8268	25.67	7.14	26.35	0.8124
	3	468.86	0.7885	433.89	0.7589	53.72	43.11	68.88	0.7890

Table 6. Comparison of Hyperparameter Tuning algorithm performance with Grid Search and Bayesian Optimisation search.

		Training		Testing		Overall Performance			
Hyperparmeter Tuning Algorithm	**MLP Algorithm**	**MSE**	**R**	**MSE**	**R**	**MAE**	**STD**	**MSE**	**R**
Grid Search	lm	3.97	0.9669	12.96	0.8467	1.60	1.70	2.34	0.9499
	br	0.57	0.9953	47.07	0.7562	1.05	2.55	2.76	0.9408
Bayessian Optimisation search	lm	5.20	0.9565	16.73	0.8355	2.29	4.90		
	br	3.52	0.9669	18.84	0.8567	3.40	4.24		

4.5. Evaluation of Proposed Neural Network Model at Different Locations

The evaluation results of the proposed neural network model at different study locations are presented as follows. Figures 9–11 show the proposed neural network model prediction with measured path loss data configured with the Levenberg–Marquardt training algorithm, two hidden-layer sizes and logsig-transig transfer function—the prediction performance of the developed neural network model in terms of R and MSE values. The R-value measures the prediction correlation between outputs (predicted loss data) and targets (actual loss data). The closeness of the R-value to 1 corresponds to a high positive correction. Otherwise, it is poorly correlated. Figures 12–14 are plotted R values between the predicted loss data and the actual loss values for sites 1 to 3 during training, validation and testing with neural networks. The R-values obtained from the plots are 0.97, 0.93 and 0.94 for site 1, 0.92, 0.93 and 0.94 for site 2 and 0.91, 0.93, 0.96, 0.94 for site 3. The performance plots in Figures 15–17 indicate that the MSE becomes smaller with the epoch number (one complete training/testing/validation cycle). The word 'epoch' is used here to mean a special hyperparameter term that defines the number of times (in terms of iteration) that the NN algorithms undergo during the entire data training duration. The error of test and validation display similar characteristics while predicting the measured loss across sites 1 to 3. Specifically, the validation MSE error shows that the proposed neural network model would not generalize well or fit the measured loss data well if trained further than 4, 8 and 8 epochs. The mean prediction error along measurement data points in sites 1, 2 and 3 are presented in Figures 18–20.

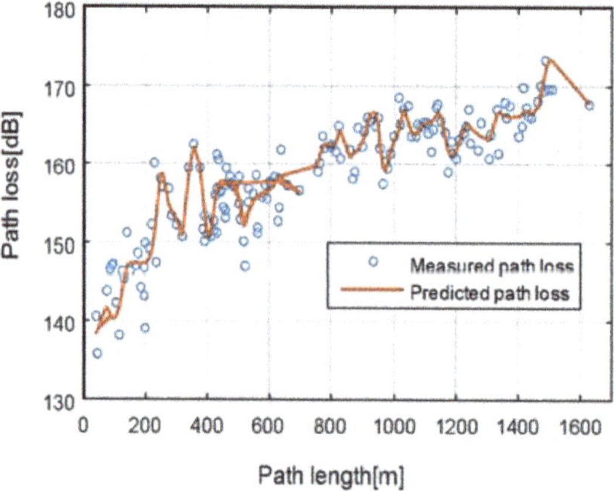

Figure 9. Comparison between measured loss and the prediction ANN model in site 1.

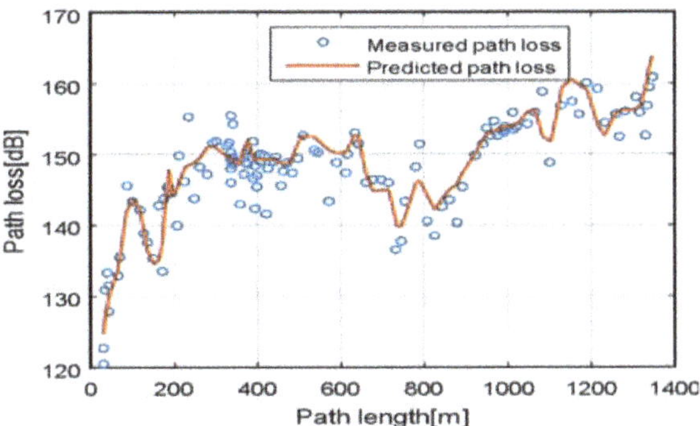

Figure 10. Comparison between measured loss and the prediction ANN model in site 2.

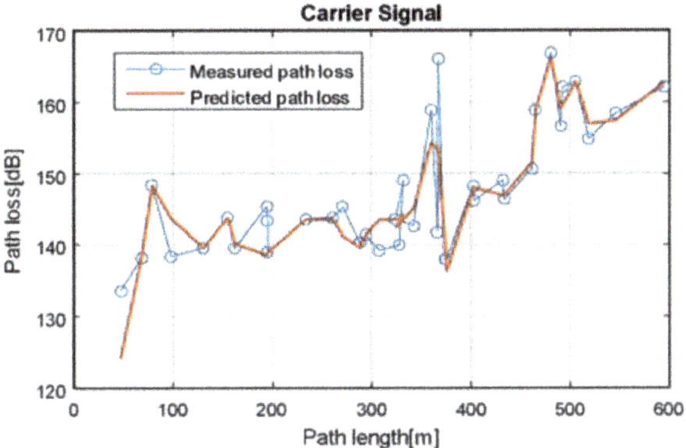

Figure 11. Comparison between measured loss and the prediction ANN model in site 3.

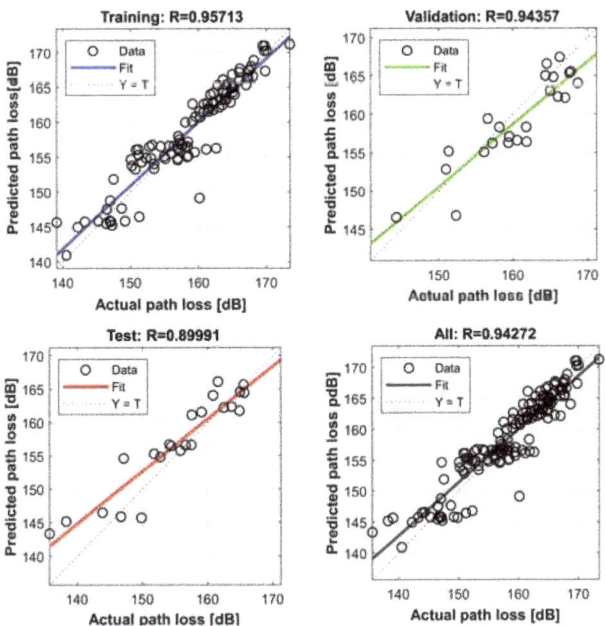

Figure 12. Prediction performance with correlation coefficient in site 1.

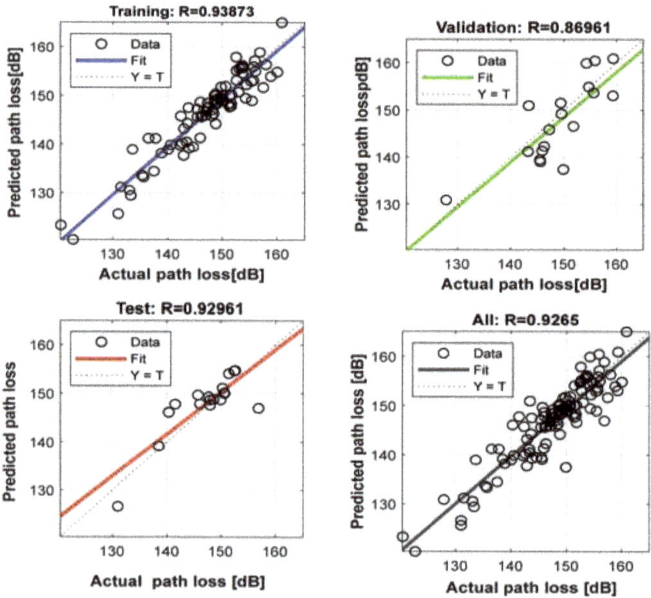

Figure 13. Prediction performance with correlation coefficient in site 2.

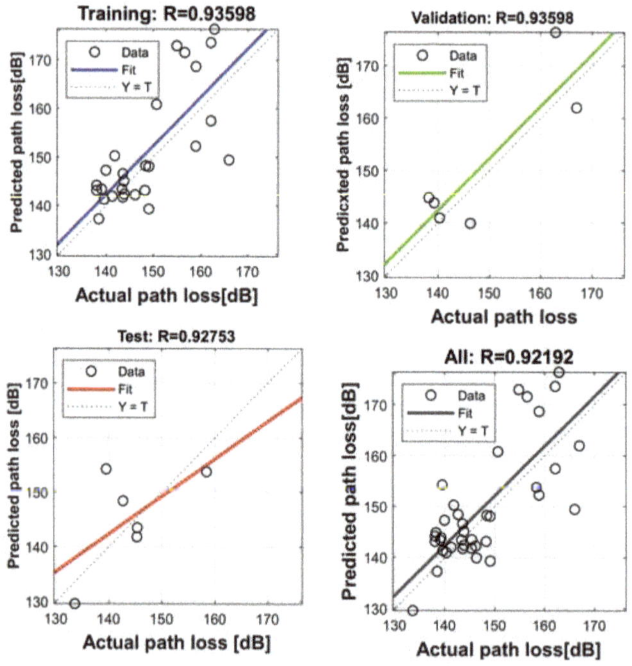

Figure 14. Prediction performance with correlation coefficient in site 3.

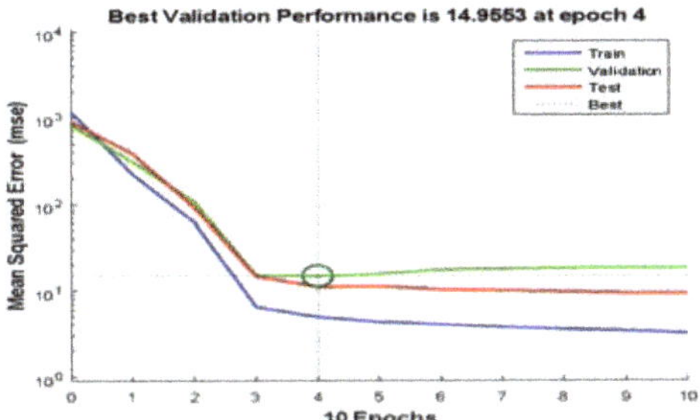

Figure 15. Network training cycles in site 1.

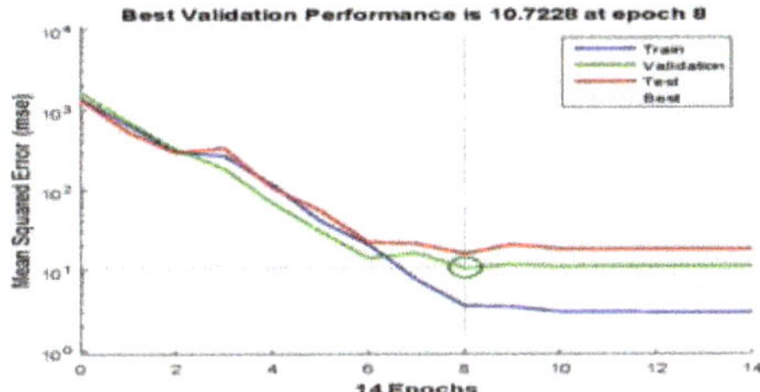

Figure 16. Network training cycles in site 2.

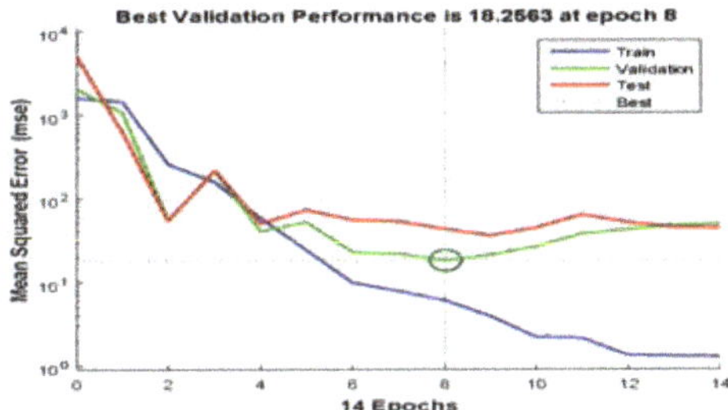

Figure 17. Network training cycles in site 3.

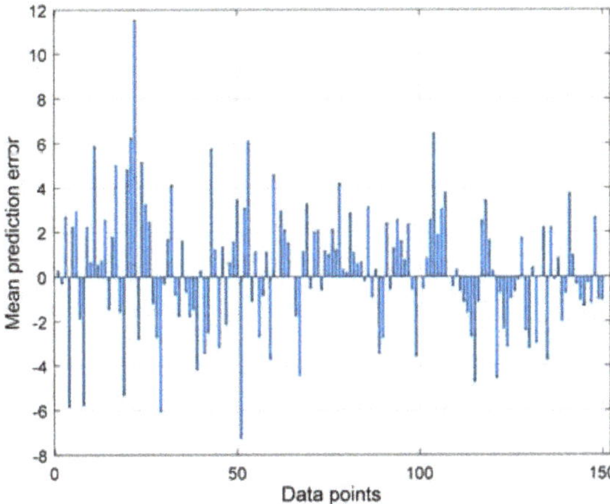

Figure 18. Mean prediction error statistics along with Data points in site 1.

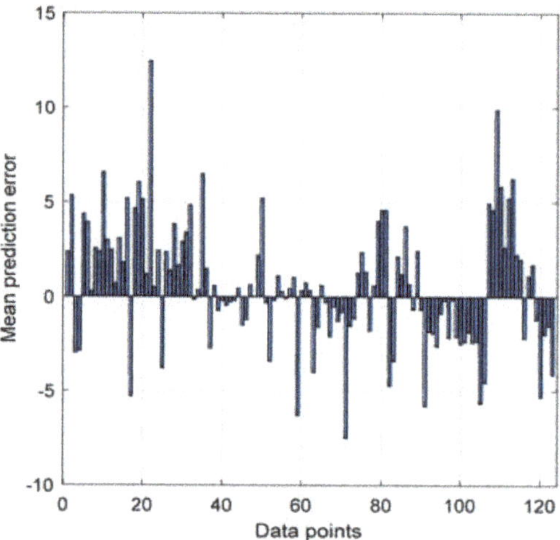

Figure 19. Mean prediction error statistics along with Data points in site 2.

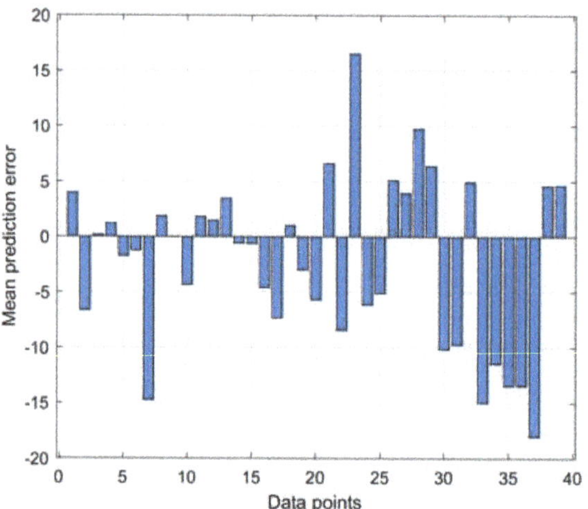

Figure 20. Mean prediction error statistics along with Data points in site 3.

4.6. Comparison of Prediction Accuracy of Proposed Neural Network Model with Log-Distance Models

Detailed prediction capabilities of all the log-distance models and the proposed model on measured path loss data are provided in the plotted graphs of Figures 21–23 in terms of MAE, RMSE and STD. The graphs show that the neural network model achieved the best predictions with marginal errors. The COST 213 (W/I) made the closest prediction to measured loss, but in terms of accuracy, the proposed neural network model achieved the best performance by 20%, 15% and 25%, respectively, across study sites. For example, while COST 213 (W/I) reached 3.34, 2.35 and 4.23 dB in terms of RMSE, the proposed neural network model attained 1.73, 2.11 and 1.45 dB across study sites. Generally, models which predict the path loss in the tested areas with RMSEs higher than the acceptable range of up to 6 dB are not selected as most suitable. However, such models could be further optimized for improved performance. The lower the RMSE value towards zero, the better the model. Regarding standard deviation error, COST 213 (W/I) achieved 1.73, 2.11 and 1.45 dB, while the proposed neural network model achieved 1.73, 2.11 and 1.45 dB, respectively. The poor predictions made by the log-distance-based models can be ascribed to due to dissimilarities and variations in environmental formations (hilly, mountainous or quasi-plain), weather conditions, soil electrical properties and terrain types that exist in different radio propagation environments.

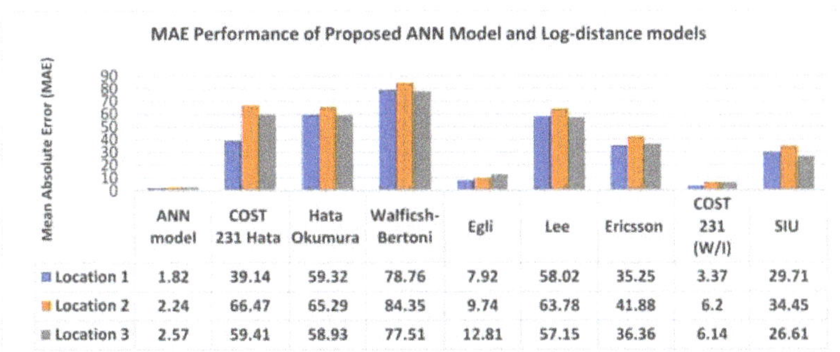

Figure 21. A comparison of mean absolute error statistics between the proposed ANN model and log-distance models on measured path loss in site locations 1, 2 and 3.

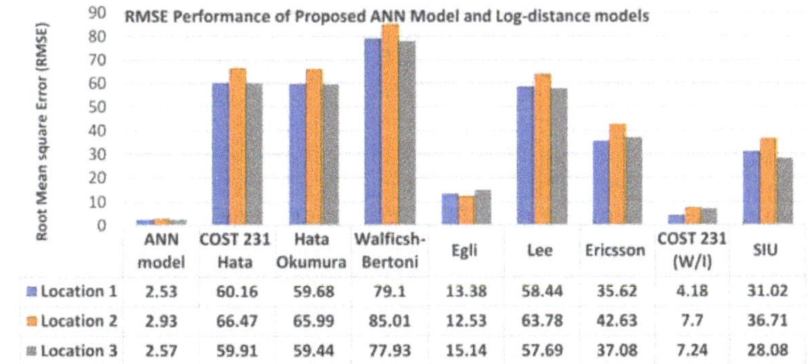

Figure 22. A comparison of root mean absolute error statistics between the proposed ANN model and log-distance models on measured path loss in site locations 1, 2 and 3.

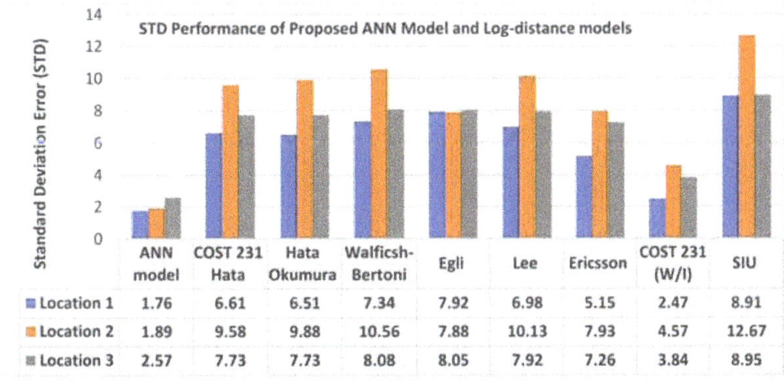

Figure 23. A comparison of standard deviation error statistics between the proposed ANN and log-distance models on measured path loss in site locations 1, 2 and 3.

5. Conclusions

The growing demand for mobile and fixed cellular telecommunication services have given substantial weight to the limited available radio frequency spectrum. Proper modelling and precise signal coverage predictions are crucial to utilizing this scarce resource effectively. Reliable predictive modeling of signal path loss aids in controlling the load on base station transmitters and assists in designing efficient radio network channels with less interference and coverage hole problems. The conventional log-distance-based statistical models for path loss prediction comprising the clustering factor, COST 234 Hata, free space, Hata, Lee models, etc., are generally limited for predicting signal attenuation losses, especially when employed in different environments other than the environment for which they have been designed.

The main objective of this paper was to develop a distinctive MLP-based path loss model with well-structured implementation network architecture, empowered with the grid search-based hyperparameter tuning method for optimal path loss approximation between mobile-station and base-station path lengths. The degree of prediction accuracy with the developed MLP network model over eight conventional log-distance-based path loss models is also clearly provided using first-order statistics. In summary, this research paper has revealed that:

- MLPANN-based path loss model with well-structured implementation network architecture, empowered with the right hyperparameter tuning algorithm is better than the standard long-distance path loss models
- the choice of both MLP-ANN modelling structure and selection of training algorithms do have a clear impact on the quality of its prediction proficiency. Specifically, in terms of MAE, RMSE and STD statistical values, the proposed model yielded up to 50% performance prediction accuracies improvement over the standard models on the acquired LTE path loss datasets.
- The selection of adaptive learning tuning hyperparameters of MLP-ANN and the tuning algorithm both have an impact on its overall predictive modelling capacity.

Future work would consider more hyperparameter selection techniques to optimize MLP model prediction accuracy during NN training. We also intend to explore more super layered training capacity of deep neural networks such as the long-short memory (LSTM) network model for predictive modelling of path loss data in our work.

Author Contributions: The manuscript was written through the contributions of all authors. J.I. was responsible for the conceptualization of the topic; article gathering and sorting were carried out by J.I., A.L.I. and S.O.; manuscript writing, original drafting and formal analysis were carried out by J.I., A.L.I., S.O., O.K., Y.K., C.-C.L. and C.-T.L.; writing of reviews and editing was carried out by J.I., A.L.I., S.O., O.K., Y.K., C.-C.L. and C.-T.L.; and J.I. led the overall research activity. All authors have read and agreed to the published version of the manuscript.

Funding: This work was supported in part by the National Research Foundation of Korea (NRF) grant funded by the Korean government (MSIT) (No. 2020R1G1A1099559).

Institutional Review Board Statement: Not applicable.

Informed Consent Statement: Not applicable.

Data Availability Statement: The data that supports the findings of this paper are available from the corresponding author upon reasonable request.

Acknowledgments: The work of Agbotiname Lucky Imoize is supported in part by the Nigerian Petroleum Technology Development Fund (PTDF) and in part by the German Academic Exchange Service (DAAD) through the Nigerian–German Postgraduate Program under grant 57473408.

Conflicts of Interest: The authors declare no conflict of interest.

References

1. Molisch, A.F. *Wireless Communications*, 2nd ed.; John Wiley & Sons, Inc.: Hoboken, NJ, USA, 2012; ISBN 9780470741870.
2. Goldsmith, A.J. *Wireless Communications*; Cambridge University Press: Cambridge, UK, 2005.
3. Isabona, J.; Imoize, A.L.; Ojo, S.; Lee, C.-C.; Li, C.-T. Atmospheric Propagation Modelling for Terrestrial Radio Frequency Communication Links in a Tropical Wet and Dry Savanna Climate. *Information* **2022**, *13*, 141. [CrossRef]
4. Isabona, J.; Konyeha, C.C.; Chinule, C.B.; Isaiah, G.P. Radio field strength propagation data and pathloss calculation methods in UMTS network. *Adv. Phys. Theor. Appl.* **2013**, *21*, 54–68.
5. Nawrocki, M.; Aghvami, H.; Dohler, M. *Understanding UMTS Radio Network Modelling, Planning and Automated Optimisation: Theory and Practice*; John Wiley & Sons: Hoboken, NJ, USA, 2006.
6. Nawrocki, M.J.; Dohler, M.; Aghvami, A.H. Modern approaches to radio network modelling and planning. In *Understanding UMTS Radio Network Modelling, Planning and Automated Optimisation: Theory and Practice*; John Wiley & Sons: Hoboken, NJ, USA, 2006; p. 3.
7. Sharma, P.K.; Singh, R.K. Comparative analysis of propagation path loss models with field measured data. *Int. J. Eng. Sci. Technol.* **2010**, *2*, 2008–2013.
8. Ajose, S.O.; Imoize, A.L. Propagation measurements and modelling at 1800 MHz in Lagos Nigeria. *Int. J. Wirel. Mob. Comput.* **2013**, *6*, 165–174. [CrossRef]
9. Ojo, S.; Imoize, A.; Alienyi, D. Radial basis function neural network path loss prediction model for LTE networks in multitransmitter signal propagation environments. *Int. J. Commun. Syst.* **2021**, *34*, e4680. [CrossRef]
10. Imoize, A.L.; Ibhaze, A.E.; Nwosu, P.O.; Ajose, S.O. Determination of Best-fit Propagation Models for Pathloss Prediction of a 4G LTE Network in Suburban and Urban Areas of Lagos, Nigeria. *West Indian J. Eng.* **2019**, *41*, 13–21.
11. Ibhaze, A.E.; Imoize, A.L.; Ajose, S.O.; John, S.N.; Ndujiuba, C.U.; Idachaba, F.E. An Empirical Propagation Model for Path Loss Prediction at 2100 MHz in a Dense Urban Environment. *Indian J. Sci. Technol.* **2017**, *10*, 1–9. [CrossRef]
12. Rathore, M.M.; Paul, A.; Rho, S.; Khan, M.; Vimal, S.; Shah, S.A. Smart traffic control: Identifying driving-violations using fog devices with vehicular cameras in smart cities. *Sustain. Cities Soc.* **2021**, *71*, 102986. [CrossRef]
13. Imoize, A.L.; Tofade, S.O.; Ughegbe, G.U.; Anyasi, F.I.; Isabona, J. Updating analysis of key performance indicators of 4G LTE network with the prediction of missing values of critical network parameters based on experimental data from a dense urban environment. *Data Br.* **2022**, *42*, 108240. [CrossRef]
14. Fujimoto, K. *Mobile Antenna Systems Handbook*; Artech House: Norwood, MA, USA, 2008; ISBN 1596931272.
15. Tataria, H.; Haneda, K.; Molisch, A.F.; Shafi, M.; Tufvesson, F. Standardization of Propagation Models for Terrestrial Cellular Systems: A Historical Perspective. *Int. J. Wirel. Inf. Networks* **2021**, *28*, 20–44. [CrossRef]
16. Hanci, B.Y.; Cavdar, I.H. Mobile radio propagation measurements and tuning the path loss model in urban areas at GSM-900 band in Istanbul-Turkey. In Proceedings of the IEEE 60th Vehicular Technology Conference, 2004, VTC2004-Fall, Los Angeles, CA, USA, 26–29 September 2004; Volume 1, pp. 139–143.
17. Ekpenyong, M.; Isabona, J.; Ekong, E. On Propagation Path Loss Models For 3-G Based Wireless Networks: A Comparative Analysis. *Comput. Sci. Telecommun.* **2010**, *25*, 74–84.
18. Isabona, J. Wavelet Generalized Regression Neural Network Approach for Robust Field Strength Prediction. *Wirel. Pers. Commun.* **2020**, *114*, 3635–3653. [CrossRef]
19. Imoize, A.L.; Oseni, A.I. Investigation and pathloss modeling of fourth generation long term evolution network along major highways in Lagos Nigeria. *Ife J. Sci.* **2019**, *21*, 39–60. [CrossRef]
20. Imoize, A.L.; Ogunfuwa, T.E. Propagation measurements of a 4G LTE network in Lagoon environment. *Niger. J. Technol. Dev.* **2019**, *16*, 1–9. [CrossRef]
21. Imoize, A.L.; Dosunmu, A.I. Path Loss Characterization of Long Term Evolution Network for. *Jordan J. Electr. Eng.* **2018**, *4*, 114–128.
22. Ekpenyong, M.E.; Robinson, S.; Isabona, J. Macrocellular propagation prediction for wireless communications in urban environments. *J. Comput. Sci. Technol.* **2010**, *10*, 130 136.
23. Nadir, Z. Empirical pathloss characterization for Oman. In Proceedings of the 2012 Computing, Communications and Applications Conference, Tamilnadu, India, 22–24 February 2012; pp. 133–137. [CrossRef]
24. Hunter, D.; Yu, H.; Pukish III, M.S.; Kolbusz, J.; Wilamowski, B.M. Selection of proper neural network sizes and architectures—A comparative study. *IEEE Trans. Ind. Informatics* **2012**, *8*, 228–240. [CrossRef]
25. Imoize, A.L.; Ibhaze, A.E.; Atayero, A.A.; Kavitha, K.V.N. Standard Propagation Channel Models for MIMO Communication Systems. *Wirel. Commun. Mob. Comput.* **2021**, *2021*, 8838792. [CrossRef]
26. Khan, N.; Haq, I.U.; Khan, S.U.; Rho, S.; Lee, M.Y.; Baik, S.W. DB-Net: A novel dilated CNN based multi-step forecasting model for power consumption in integrated local energy systems. *Int. J. Electr. Power Energy Syst.* **2021**, *133*, 107023. [CrossRef]
27. Jo, H.-S.; Park, C.; Lee, E.; Choi, H.K.; Park, J. Path loss prediction based on machine learning techniques: Principal component analysis, artificial neural network, and Gaussian process. *Sensors* **2020**, *20*, 1927. [CrossRef]
28. Guo, D.; Zhang, Y.; Xiang, Q.; Li, Z. Improved radio frequency identification indoor localization method via radial basis function neural network. *Math. Probl. Eng.* **2014**, *2014*, 420482. [CrossRef]

29. Guo, Y.; Liu, Y.; Li, S. Modeling and Simulation of Terahertz Indoor Wireless Channel Based on Radial Basis Function Neural Network. In Proceedings of the 2021 International Conference on Microwave and Millimeter Wave Technology (ICMMT), Nanjing, China, 23–26 May 2021; pp. 1–3.
30. Annepu, V.; Rajesh, A.; Bagadi, K. Radial basis function-based node localization for unmanned aerial vehicle-assisted 5G wireless sensor networks. *Neural Comput. Appl.* **2021**, *33*, 12333–12346. [CrossRef]
31. Huang, G.-B. Learning capability and storage capacity of two-hidden-layer feedforward networks. *IEEE Trans. Neural Netw.* **2003**, *14*, 274–281. [CrossRef]
32. Abhayawardhana, V.S.; Wassellt, I.J.; Crosby, D.; Sellars, M.P.; Brown, M.G. Comparison of empirical propagation path loss models for fixed wireless access systems. In Proceedings of the IEEE Vehicular Technology Conference, Stockholm, Sweden, 30 May–1 June 2005; Volume 61, pp. 73–77.
33. Hinga, S.K.; Atayero, A.A. Deterministic 5G mmWave Large-Scale 3D Path Loss Model for Lagos Island, Nigeria. *IEEE Access* **2021**, *9*, 134270–134288. [CrossRef]
34. Haykin, S. Neural networks: A guided tour. *Soft Comput. Intell. Syst. theory Appl.* **1999**, *71*, 71–80.
35. Isabona, J.; Osaigbovo, A.I. Investigating predictive capabilities of RBFNN, MLPNN and GRNN models for LTE cellular network radio signal power datasets. *FUOYE J. Eng. Technol.* **2019**, *4*, 155–159. [CrossRef]
36. Sotiroudis, S.P.; Goudos, S.K.; Gotsis, K.A.; Siakavara, K.; Sahalos, J.N. Modeling by optimal artificial neural networks the prediction of propagation path loss in urban environments. In Proceedings of the 2013 IEEE-APS Topical Conference on Antennas and Propagation in Wireless Communications (APWC), Turin, Italy, 9–13 September 2013; pp. 599–602.
37. Isabona, J.; Imoize, A.L. Terrain-based adaption of propagation model loss parameters using non-linear square regression. *J. Eng. Appl. Sci.* **2021**, *68*, 33. [CrossRef]
38. Phillips, C.; Sicker, D.; Grunwald, D. A Survey of Wireless Path Loss Prediction and Coverage Mapping Methods. *IEEE Commun. Surv. Tutorials* **2013**, *15*, 255–270. [CrossRef]
39. Imoize, A.L.; Oyedare, T.; Ezekafor, C.G.; Shetty, S. Deployment of an Energy Efficient Routing Protocol for Wireless Sensor Networks Operating in a Resource Constrained Environment. *Trans. Networks Commun.* **2019**, *7*, 34–50. [CrossRef]
40. Imoize, A.L.; Ajibola, O.A.; Oyedare, T.R.; Ogbebor, J.O.; Ajose, S.O. Development of an Energy-Efficient Wireless Sensor Network Model for Perimeter Surveillance. *Int. J. Electr. Eng. Appl. Sci.* **2021**, *4*, 1–17.
41. Imoize, A.L. Analysis of Propagation Models for Mobile Radio Reception at 1800MHz. Master's Thesis, University of Lagos, Lagos, Nigeria, 2011.
42. Isabona, J.; Ojuh, D. Adaptation of Propagation Model Parameters toward Efficient Cellular Network Planning using Robust LAD Algorithm. *Int. J. Wirel. Microw. Technol.* **2020**, *10*, 13–24. [CrossRef]
43. Imoize, A.L.; Orolu, K.; Atayero, A.A.-A. Analysis of key performance indicators of a 4G LTE network based on experimental data obtained from a densely populated smart city. *Data Br.* **2020**, *29*, 105304. [CrossRef] [PubMed]
44. Ojo, S.; Akkaya, M.; Sopuru, J.C. An ensemble machine learning approach for enhanced path loss predictions for 4G LTE wireless networks. *Int. J. Commun. Syst.* **2022**, *35*, e5101. [CrossRef]
45. Ebhota, V.C.; Isabona, J.; Srivastava, V.M. Improved adaptive signal power loss prediction using combined vector statistics based smoothing and neural network approach. *Prog. Electromagn. Res. C* **2018**, *82*, 155–169. [CrossRef]
46. Ebhota, V.C.; Isabona, J.; Srivastava, V.M. Base line knowledge on propagation modelling and prediction techniques in wireless communication networks. *J. Eng. Appl. Sci.* **2018**, *13*, 1919–1934.
47. Coskun, N.; Yildirim, T. The effects of training algorithms in MLP network on image classification. In Proceedings of the International Joint Conference on Neural Networks, Portland, OR, USA, 20–24 July 2003; Volume 2, pp. 1223–1226.
48. Salman, M.A.; Popoola, S.I.; Faruk, N.; Surajudeen-Bakinde, N.T.; Oloyede, A.A.; Olawoyin, L.A. Adaptive Neuro-Fuzzy model for path loss prediction in the VHF band. In Proceedings of the 2017 International Conference on Computing Networking and Informatics (ICCNI), Lagos, Nigeria, 29–31 October 2017; pp. 1–6.
49. Aldossari, S.; Chen, K.-C. Predicting the path loss of wireless channel models using machine learning techniques in mmwave urban communications. In Proceedings of the 2019 22nd International Symposium on Wireless Personal Multimedia Communications (WPMC), Lisbon, Portugal, 24–27 November 2019; pp. 1–6.
50. Ahmadien, O.; Ates, H.F.; Baykas, T.; Gunturk, B.K. Predicting Path Loss Distribution of an Area From Satellite Images Using Deep Learning. *IEEE Access* **2020**, *8*, 64982–64991. [CrossRef]
51. Turan, B.; Coleri, S. Machine learning based channel modeling for vehicular visible light communication. *IEEE Trans. Veh. Technol.* **2021**, *70*, 9659–9672. [CrossRef]
52. Ates, H.F.; Hashir, S.M.; Baykas, T.; Gunturk, B.K. Path loss exponent and shadowing factor prediction from satellite images using deep learning. *IEEE Access* **2019**, *7*, 101366–101375. [CrossRef]
53. Huang, J.; Cao, Y.; Raimundo, X.; Cheema, A.; Salous, S. Rain statistics investigation and rain attenuation modeling for millimeter wave short-range fixed links. *IEEE Access* **2019**, *7*, 156110–156120. [CrossRef]
54. Zhang, Y.; Wen, J.; Yang, G.; He, Z.; Wang, J. Path loss prediction based on machine learning: Principle, method, and data expansion. *Appl. Sci.* **2019**, *9*, 1908. [CrossRef]
55. Nguyen, D.C.; Cheng, P.; Ding, M.; Lopez-Perez, D.; Pathirana, P.N.; Li, J.; Seneviratne, A.; Li, Y.; Poor, H.V. Enabling AI in future wireless networks: A data life cycle perspective. *IEEE Commun. Surv. Tutor.* **2020**, *23*, 553–595. [CrossRef]

56. Ferreira, G.P.; Matos, L.J.; Silva, J.M.M. Improvement of outdoor signal strength prediction in UHF band by artificial neural network. *IEEE Trans. Antennas Propag.* **2016**, *64*, 5404–5410. [CrossRef]
57. Singh, H.; Gupta, S.; Dhawan, C.; Mishra, A. Path loss prediction in smart campus environment: Machine learning-based approaches. In Proceedings of the 2020 IEEE 91st Vehicular Technology Conference (VTC2020-Spring), Antwerp, Belgium, 25–28 May 2020; pp. 1–5.
58. Thrane, J.; Zibar, D.; Christiansen, H.L. Model-aided deep learning method for path loss prediction in mobile communication systems at 2.6 GHz. *IEEE Access* **2020**, *8*, 7925–7936. [CrossRef]
59. Shrestha, A.; Mahmood, A. Review of deep learning algorithms and architectures. *IEEE Access* **2019**, *7*, 53040–53065. [CrossRef]
60. Nguyen, C.; Cheema, A.A. A deep neural network-based multi-frequency path loss prediction model from 0.8 GHz to 70 GHz. *Sensors* **2021**, *21*, 5100. [CrossRef]
61. Challita, U.; Dong, L.; Saad, W. Deep learning for proactive resource allocation in LTE-U networks. In Proceedings of the European Wireless Technology Conference, Dresden, Germany, 15 May 2017.
62. Song, W.; Zeng, F.; Hu, J.; Wang, Z.; Mao, X. An unsupervised-learning-based method for multi-hop wireless broadcast relay selection in urban vehicular networks. In Proceedings of the 2017 IEEE 85th Vehicular Technology Conference (VTC Spring), Sydney, NSW, Australia, 4–7 June 2017; pp. 1–5.
63. Ebhota, V.C.; Isabona, J.; Srivastava, V.M. Environment-Adaptation Based Hybrid Neural Network Predictor for Signal Propagation Loss Prediction in Cluttered and Open Urban Microcells. *Wirel. Pers. Commun.* **2019**, *104*, 935–948. [CrossRef]
64. Isabona, J.; Babalola, M. Statistical Tuning of Walfisch-Bertoni Pathloss Model based on Building and Street Geometry Parameters in Built-up Terrains. *Am. J. Phys. Appl.* **2013**, *1*, 10–16.
65. Bird, J. *Engineering Mathematics*, 5th ed.; Newness: Los Angeles, CA, USA, 2010; ISBN 9780750681520.

Article

Patch-Wise Infrared and Visible Image Fusion Using Spatial Adaptive Weights

Syeda Minahil †, Jun-Hyung Kim † and Youngbae Hwang *

Department of Intelligent Systems and Robotics, Chungbuk National University, Cheongju 28644, Korea; SyedaMinahil@outlook.com (S.M.); mohl@naver.com (J.-H.K.)
* Correspondence: ybhwang@cbnu.ac.kr
† These authors contributed equally to this work.

Abstract: In infrared (IR) and visible image fusion, the significant information is extracted from each source image and integrated into a single image with comprehensive data. We observe that the salient regions in the infrared image contain targets of interests. Therefore, we enforce spatial adaptive weights derived from the infrared images. In this paper, a Generative Adversarial Network (GAN)-based fusion method is proposed for infrared and visible image fusion. Based on the end-to-end network structure with dual discriminators, a patch-wise discrimination is applied to reduce blurry artifact from the previous image-level approaches. A new loss function is also proposed to use constructed weight maps which direct the adversarial training of GAN in a manner such that the informative regions of the infrared images are preserved. Experiments are performed on the two datasets and ablation studies are also conducted. The qualitative and quantitative analysis shows that we achieve competitive results compared to the existing fusion methods.

Keywords: infrared and visible image fusion; Generative Adversarial Network; patchGAN; dual-discriminator; spatial adaptive weights

Citation: Minahil, S.; Kim, J.-H.; Hwang, Y. Patch-Wise Infrared and Visible Image Fusion Using Spatial Adaptive Weights. *Appl. Sci.* **2021**, *11*, 9255. https://doi.org/10.3390/app11199255

Academic Editor: Pavel Lyakhov

Received: 14 September 2021
Accepted: 30 September 2021
Published: 5 October 2021

1. Introduction

In practical applications, a fused image is essential to contain high-quality details for attaining a comprehensive representation of the real scene [1]. Nowadays, various image fusion methods have been proposed and are usually divided into several categories that include sparse representation-based methods [2,3], gradient-based methods [4], wavelet transformation-based methods [5,6], neural network-based methods [7] and deep learning based methods [8,9]. Deep learning based infrared and visible image fusion methods exploit the features of images of different modalities and integrate them into one single image with the composite information. The infrared images reflect the thermal radiation of objects [10] and visible images contain the textural information. Infrared and visible image fusion methods [1,9,11–13] extract the characteristics of both the images and achieve images that improve visual understanding and are beneficial in various fields like computer vision in object detection, recognition and military surveillance [10].

The recent fusion algorithms achieve promising results. Guided filter [14] is widely used for the purpose of image fusion which involves two-scale decomposition of the image. Then, the base layer (containing large scale variations in intensity) and detail layer (capturing small scale details) are fused together using a guided filtering based weighted average method. In Deepfuse [15], there is a Siamese based encoder with two CNN layers that extracts the features of the source images. These maps are fused by the addition strategy. The decoding network with three CNN layers reconstructs the fused image. Although it achieves better results, the network is too simple and cannot extract the salient features properly. In Densefuse [9], Ma et al. proposed a method based on dense block and an auto-encoder module. Denseblock has skip connections which help to preserve more features. The drawback of this approach is that the fusion is not considered in the training process and only the auto-encoder is trained for the reconstruction of the images. The CNN based methods usually rely on the ground truth and in case of fusion of infrared and visible images there are no predefined standards [16]. Generative Adversarial Network (GAN)-based fusion methods are also very popular recently. These methods use traditional GAN or its variants for fusing the images, but the trained image is trained to be more similar to only one of the source images which catalyze the loss of information existing in the other image. In DDcGAN [12], GAN is applied with dual discriminators where each discriminator is tasked to make the fused image more similar to the infrared and visible image, respectively, without the ground truth.

In this paper, we apply a patch-wise discriminator inspired from the patchGAN [17] in the base structure using dual discriminators [12]. Vanilla GAN validates the authenticity of the entire generated image while the patchGAN verifies the authenticity in units of patches of $N \times N$ size of an image. By doing so, an image is only considered as real if all the patches attain a high probability of being real. This also reduces the computation speed, the number of parameters and is unaffected by the size of the images. In our approach, we also assign adaptive weights to IR image based on the observation that in the IR image only the high activation regions contains the salient information while the remaining background region contains very less or no information. We utilize these weight maps in the loss function.

To summarize, this paper makes the following contributions;

1. We propose a new framework for infrared and visible image fusion in which the patchGAN is applied to the dual discriminator network structure.
2. We introduce a new loss function based on constructing adaptive weight maps based on the IR image to preserve only the important information from both the infrared and visible images.
3. Our method produces competitive results as compared with the existing fusion methods quantitatively as well as qualitatively.

The rest of our paper is structured as follows: related works are briefly reviewed in Section 2. In Section 3, we present our proposed method in detail. In Section 4, we illustrate our experimental results and ablation studies. Finally, the conclusions are drawn in Section 5.

2. Related Works

Generative adversarial networks (GANs) [18] are one of the generative models. GANs have achieved impressive success in generating images from existing images or random noise. In GANs, we have a generator which is purposed to generate real-like fake sampled images with adversarial loss which steers the output image to be indistinguishable from the real images and to be able to fool the discriminator. The discriminator behaves as a classifier and provides the probability of the data being real or fake. The training process of a generator and a discriminator forms an adversarial process and is continued until the discriminator is unable to distinguish the generated samples. The training of GAN is a critical task because of its unstable training. To alleviate this, GANs with a conditional settings were proposed. In conditional GANs (cGAN) [19], the generator and discriminator are conditioned on some auxiliary information. This additional information is generally labeled data. This provides the guidelines to the generator in figuring out what kind of data needs to be generated. Generally, GANs use a cross entropy loss function which may lead to the vanishing gradient problem during training. To overcome this issue, Least Squares Generative Adversarial Networks (LSGANs) [20], which use least square loss for the discriminator, and WGANs [21], which use the Wasserstein loss were proposed.

The first endeavour of utilizing GAN for the task of infrared and visible image fusion was proposed in FusionGAN [22] in which the fused image is compelled to contain more texture details by introducing the discriminator to distinguish the fused image from the visible image . In GAN-based image fusion methods, the discriminator makes the generated image more similar to the visible image and this problem is alleviated in the DDcGAN [12] by the introduction of two discriminators. The generator makes the fused image and one discriminator distinguishes the fused image from the visible image while the other discriminator distinguishes the fused image from the infrared image. The generator consists of two deconvolution layers and an auto-encoder network. The visible and a low resolution infrared images are passed through the deconvolution layers to get the same resolution. The output of the deconvolution layers are concatenated together and fed into the encoder for the feature extraction and fusion process. The fused feature maps are given to the decoder for the reconstruction of fused image which has the same resolution as the visible image. The encoder has a DenseNet consisting of short connections which improve the feature extraction. Both the discriminators have the same architecture and play an adversarial role against the generator. In DDcGAN, the discriminators are not only supposed to contemplate their adversarial role but also maintain the balance between both the discriminators. The loss of generator is the summation of adversarial loss and the weighted content loss. This adversarial loss comes from the discriminators.

In U2Fusion [23], an end-to-end semi-supervised fusion method is proposed which can fuse multi-focus, multi-exposure and multi-modal images. This method automatically approximates the significance of the source images and suggests and adaptive information preservation degree. This adaptive information preservation degree is utilized to conserve the similarity between the fusion result and source images. NestFuse [24] inspired by the DenseFuse, preserves more multi-scale features from the visible image while enhancing the salient features of infrared images. Their model is based on nest connections and Spatial/Channel attention models. The spatial attention models signifies the importance of each spatial position whereas the channel attention models uses deep features.

In PatchGAN [17], the conditional GANs are used for image-to-image translation. The architecture of the generator and the discriminators differ from the previous works utilizing GANs for the same task. The generator used has a U-Net based architecture and the discriminator is a markovian discriminator [17,25]. PatchGAN does not signify the whole image as fake or real instead of evaluating the local patches from the images. The patch size can be adjusted by changing the size of the kernel in convolution layers or the number of layers in the discriminator. In [17], an input image of 256×256 size is concatenated with the generated 256×256 image and provided to the generator G. The patch size used in this paper is 70×70. The generator provides a feature map of $30 \times 30 \times 1$ which means that each pixel of this map corresponds to the 70×70 patch of the input image. All the values of the $30 \times 30 \times 1$ feature maps are averaged to estimate the probability of the patch being real or fake. This makes patchGAN more attentive to the local features of the images. PatchGAN is now being widely used in many applications. In PGGAN [26], the patchGAN is combined with the globalGAN for the task of image inpainting. The discriminator of PGGAN, first uses the shared layers between the patchGAN and the global GAN to learn the basic low-level features which is later split to generate two separate adversarial losses to preserve both the local and the global features in images. In [27], the author proposes an image text deblurring method using two GAN networks which are used to convert the blurred images to deblurred images and the deblurred images to blurred images which helps in putting the constraints on the generated samples. The discriminator used in this model is patchGAN discriminator. PatchGAN is also used in multilevel feature fusion for underwater image color correction [28]. In this model, the multi-scale features are extracted and then global features are fused together with low-level features at each scale.

3. Proposed Method

In this section, we first elaborate our proposed end-to-end deep learning based fusion network, then we discuss the formation of weight map. At last, we introduce the design of our new loss function.

In our fusion network, we apply patchGAN [17] in the dual discriminator structure framework [12], as shown in Figure 1. Our model has one generator G and two discriminators D_i and D_v. Given the infrared image i and visible image v, the task of the generator is to generate a fused image which should be able to fool the discriminators. D_v aims to distinguish the generated image from the visible image, while D_i is trained to discriminate between the original infrared image and the fused image. In general, the output of the discriminators is a scalar value that approximates the probability of the input from the source data rather than generated data G.

Inspired by the patchGAN [17], the generator as well as the discriminator is of the form convolution-BatchNorm-ReLu [12,29,30]. The generator's architecture is based on U-Net [31] with the auto-encoder system with skip connections. The discriminators used for our model are Markovian discriminators [17,25], "PatchGAN", that only penalize structure at the scale of patches. This discriminator tries to signify if each $N \times N$ patch in an image is real or fake. In vanilla GAN, given an input image, the output is a single probability of the image being real or fake, but here we get an $N \times N$ array of output X where each X_{ij} represents the probability of each patch of an image being real or fake.

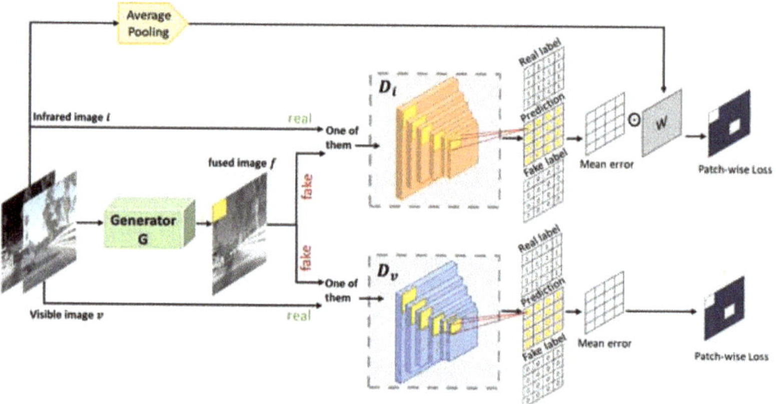

Figure 1. The overall architecture of our proposed method.

We have observed that in infrared image, only the high activation region contains significant information. The complexity in IR-visible image fusion research lies in correctly extracting the information on thermal targets from the IR image, and trying to keep obvious background information of the visible image in the fusion image [32]. We intend to give more weight to the informative regions of IR. For this purpose, we generate an adaptive weight map W based on the IR image and utilize this weight map in the adversarial loss during training in a fashion which could preserve the informative region of the IR image. For the construction of the weight map W, we take the IR image and apply average pooling. We can choose average pooling as well as max-pooling strategy to create the weight maps equal to the size of the output of the discriminator. The reason to choose average pooling is that the output of the average pooling can be considered as a smooth version of the input image. In this way, more weights can be given to high activation regions of the input infrared image. In contrast, max pooling can cause an abrupt change in weights of adjacent pixels. We have also compared these two strategies in ablation studies.The size of the weight map should match the size of the output of the discriminator.

Loss Function

Loss function plays a principle role in the learning of any model. In usual GAN, the discriminator is trained on the real and the generated images but the generator is trained indirectly via discriminator. In infrared and visible image fusion methods, the dual purposed generator is not only tasked to fool the discriminator but it should also keep the correspondence between the generated image and the source images. This is supervised by the loss function of G. Each discriminator is first trained with the input patches (i.e., the visible image patches in D_v and the infrared image patches in D_i) and then the fused patches. The total loss of each discriminator is the sum of both the losses which aims to discriminate between the fused image by G and the source images.

For the discriminators, the patches from the source images are real data while the patches from the generated/fused images are fake data. The discriminator D_i is first trained on the loss between the patch from input infrared image and the real label vector with each value 1. Then, this D_i is trained on the loss between the patch from fused image and the fake label vector with each value 0. Similarly, the discriminator D_v learns the loss between the patch from input visible image and the real label vector. After this, this D_v is trained on the loss between the patch from fused image and the fake label vector. We are using the mean square error loss for the discriminators.

Our purpose is to train the network in such a way that it gives more weight to the high activation region of infrared images which contains the significant information. For this, we define a new loss function for each patch by infusing the weights in the loss. We propose two methods of doing this; one is to use this new weighted loss only in the discriminator D_i as $L^*_{D_i}$, and the other method is to use it in the discriminator D_i as $L^*_{D_i}$ and generator as $L^{adv*}_{D_i}$ simultaneously. Generally, the loss of the discriminator is defined as follows:

$$L_D = \min_D V(D) = \frac{1}{2}\mathbb{E}[(D(x) - a)^2] + \frac{1}{2}\mathbb{E}[(D(G(x)) - b)^2] \tag{1}$$

where $D(x)$ and $D(G(x))$ is the output of discriminator given the image x and the generated image $G(x)$, respectively . 'a' is the target label vector which is 1 in the case of source images and 'b' is the target label vector which is 0 in the case of a fused image.

Using the new loss in D_i, the losses of both the discriminators become

$$L^*_{D_i} = \min_D V(D_i) = \frac{1}{2}\mathbb{E}[W * (D_i(i) - a)^2] + \frac{1}{2}\mathbb{E}[W * (D_i(f) - b)^2] \tag{2}$$

$$L_{D_v} = \min_D V(D_v) = \frac{1}{2}\mathbb{E}[(D_v(v) - a)^2] + \frac{1}{2}\mathbb{E}[(D_v(f) - b)^2] \tag{3}$$

The loss function of the generator consists of the content loss and the adversarial loss:

$$L_G = L_{con} + L_{adv} \tag{4}$$

L_{ssim} and L_{mse} are the structural similarity loss and the mean square error loss, respectively, between the input images and the generated/fused image. They are used as the content loss.

$$L_{con} = L_{ssim} + L_{mse} \tag{5}$$

$L^{adv}_{D_i}$ and $L^{adv}_{D_v}$ are the adversarial losses provided by the discriminators D_i and D_v. Here, mean square error loss is used as an adversarial loss. So the total loss of G becomes

$$L_{Total} = L_{ssim} + L_{mse} + \gamma L^{adv}_{D_i} + \lambda L^{adv}_{D_v} \tag{6}$$

In the case of using the new loss in the generator, we replace the mean square error loss $L^{adv}_{D_i}$ with the new loss $L^{adv*}_{D_i}$. Here, the output of the discriminator is compared with real label only.

$$L^{adv*}_{D_i} = \min_G V(G) = \frac{1}{2}\mathbb{E}[W * (D_i - a)^2] \tag{7}$$

$$L^{adv}_{D_v} = \min_G V(G) = \frac{1}{2}\mathbb{E}[(D_v - a)^2] \tag{8}$$

4. Experiments

In this section, we describe our experimental results. We have conducted extensive evaluation and comparison study against state-of-the-art algorithms including U2Fusion [23],

NestFuse [24] and DDcGAN [12]. We first conduct our experiments on the images taken from the RoadScene [23] and our private dataset. In the original patchGAN paper [17], they used an input image of size 256 × 256 and a patch size of 70 × 70. However, for our experiments, for the input image of 512 × 256, we have taken a patch size of 65 × 65 and a learning rate of 0.0001. We also change the values of γ, but for the qualitative analysis both λ and γ are fixed at 0.5. For the quantitative comparison, we select 20 images with different conditions, including indoor and outdoor, and day and night. For the verification of our results, we choose six quality metrics; Correlation coefficient (CC) that measures the degree of linear correlation between the source images and the fused image, sum of correlation of differences (SCD) [33], FMI_{dct} [34] calculates the mutual information for discrete cosine feature, modified Structural Similarity ($SSIM_a$) and multi scale SSIM (MSSSIM) for no reference image which models the loss and distortion between two images according to their similarities in light, contrast and structure information and Peak signal-to-noise ratio (PSNR).

4.1. Qualitative Analysis

The fused images attained by the three state-of-the-art algorithms and our proposed method are shown in Figures 2 and 3. We analyze the relative performance on three images from RoadScene and three from our private dataset using the new loss in discriminator $L^*_{D_i}$ as well as generator $L^{adv*}_{D_i}$.

We can see that images created by our method preserve the thermal targets from the infrared images as in the third example of Figure 2. It can also preserve more textural information from visible images such as sky (red box) in the first example of Figure 3. If we look at the overall images created by these methods, we would observe that the DDcGAN creates blurry images. NestFuse gives better results in first example of Figure 3, but some salient features are not clear such as in the red box in second image of Figure 3. Generally our proposed method tries to preserve as much information from both the infrared as well as the visible images and tries to maintain the overall good quality of images, simultaneously visible in the third and the fourth images.

4.2. Quantitative Analysis

For the quantitative comparison, we take the average value of 20 fused images for each metric. In this comparison, we analyze the new weighted loss in discriminator D_{ir} and generator G, and also in discriminator D_{ir} only. In Tables 1 and 2, we have taken fixed values of $\lambda = 0.5$ and $\gamma = 0.1$. Table 1 displays the values of different metrics for RoadScene dataset while Table 2 displays the values for our private dataset. The best values are indicated in red, the second best values are indicated in blue, and the third highest values in cyan.

From Table 1, we can see that as compared to the state-of-the-art methods, our method achieves the top-2 results in CC and $SSIM_a$. The second and third best results in MS-SSIM, SCD and PSNR. The second best result has a narrow margin of only 0.018 from the best result in PSNR, a difference of only 0.0252 from the best result in MS-SSIM and a difference of 0.0386 from the best result in SCD.

In Table 2, as compared to the other state-of-the-art methods, our method acquires the top-2 ranks in MS-SSIM, FMI_{dct} which calculates the mutual information for discrete cosine features and $SSIM_a$. The second best and the third best results in CC, SCD and PSNR. CC achieves the second best result with a difference of 0.0017 from the best result. SCD has a difference of only 0.0002 and PSNR has a margin of 0.16 from the best results. The highest values in FMI_{dct} and $SSIM_a$ indicate that our method attains more features and structural information. Highest value in PSNR indicates that the fused image is more similar to the source images and is of higher quality with less distortion. In general, our method gives better performance than DDcGAN by simply replacing the GAN with PatchGAN and adding weights. These results prove the effectiveness of our method.

Figure 2. Qualitative comparison of the proposed method with other state-of-the-art methods on image pairs taken from the private dataset. The first row contains visible images. Second row contains infrared images. The last row contains the fused results of our proposed method.

We also compare the different values of γ in Tables 3 and 4. Here, the best values are indicated in red and the second best values are indicated in blue. From Tables 3 and 4, we can witness that most of the highest results are achieved with $\lambda = 0.5$ and $\gamma = 0.1$.

Figure 3. Qualitative comparison of the proposed method with other state-of-the-art methods on image pairs taken from the RoadScene dataset. The first row contains visible images. Second row contains infrared images. The last row contains the fused results of our proposed method.

Table 1. The average values of quality metrics for 20 fused images of our RoadScene dataset with $\lambda = 0.5$ and $\gamma = 0.1$ used in the loss function. The best values are indicated in red, the second best values are indicated in blue and the third highest values in cyan.

Methods		CC	$MSSSIM$	SCD [33]	FMI_{dct} [34]	$SSIM_a$	PSNR
U2Fusion [23]		1.2199	0.8907	1.3236	0.3057	0.7204	16.0903
NestFuse [24]		1.1889	0.8376	1.6574	0.2969	0.6598	13.8161
DDcGAN [12]		1.1752	0.7067	1.5041	0.3589	0.5965	13.8019
ours ($L^{*}_{D_i}$, $L^{adv*}_{D_i}$)	$\lambda = 0.5, \gamma = 0.1$	1.2675	0.8655	1.6188	0.2691	0.7381	16.0222
ours ($L^{adv*}_{D_i}$)	$\lambda = 0.5, \gamma = 0.1$	1.2587	0.8625	1.5609	0.2736	0.7396	16.0723

Table 2. The average values of quality metrics for 20 fused images of our private dataset with $\lambda = 0.5$ and $\gamma = 0.1$ used in the loss function. The best values are indicated in red, the second best values are indicated in blue and the third highest values in cyan.

Methods		CC	$MSSSIM$	SCD [33]	FMI_{dct} [34]	$SSIM_a$	PSNR
U2Fusion [23]		1.3785	0.8954	0.9211	0.1953	0.7426	19.2835
NestFuse [24]		1.4732	0.899	1.4616	0.2369	0.7322	18.7873
DDcGAN [12]		1.2685	0.7785	1.1741	0.1879	0.5409	11.5284
ours ($L^*_{D_i}$, $L^{adv*}_{D_i}$)	$\lambda = 0.5, \gamma = 0.1$	1.4715	0.9068	1.4614	0.2684	0.7594	19.0826
ours ($L^{adv*}_{D_i}$)	$\lambda = 0.5, \gamma = 0.1$	1.4701	0.905	1.3959	0.2714	0.7541	19.1149

Table 3. The average values of quality metrics for 20 fused images of our RoadScene dataset. Different values of λ and γ are used in the loss function. The best values are indicated in red and the second best values are indicated in blue.

Methods		CC	$MSSSIM$	SCD [33]	FMI_{dct} [34]	$SSIM_a$	PSNR
ours ($L^*_{D_i}$, $L^{adv*}_{D_i}$)	$\lambda = 0.5, \gamma = 0.1$	1.2675	0.8655	1.6188	0.2691	0.7381	16.0222
ours ($L^*_{D_i}$, $L^{adv*}_{D_i}$)	$\lambda = 0.5, \gamma = 0.5$	1.2549	0.8785	1.6598	0.2501	0.7319	15.8065
ours ($L^*_{D_i}$, $L^{adv*}_{D_i}$)	$\lambda = 0.5, \gamma = 5$	1.2379	0.8601	1.6986	0.2346	0.7139	15.2591
ours ($L^{adv*}_{D_i}$)	$\lambda = 0.5, \gamma = 0.1$	1.2587	0.8625	1.5609	0.2736	0.7396	16.0723
ours ($L^{adv*}_{D_i}$)	$\lambda = 0.5, \gamma = 0.5$	1.2567	0.8764	1.6566	0.2569	0.7326	15.8283
ours ($L^{adv*}_{D_i}$)	$\lambda = 0.5, \gamma = 5$	1.2165	0.8547	1.6916	0.24	0.6998	14.7879

Table 4. The average values of quality metrics for 20 fused images of our private dataset. Different values of λ and γ are used in the loss function. The best values are indicated in red and the second best values are indicated in blue.

Methods		CC	$MSSSIM$	SCD [33]	FMI_{dct} [34]	$SSIM_a$	PSNR
ours ($L^*_{D_i}$, $L^{adv*}_{D_i}$)	$\lambda = 0.5, \gamma = 0.1$	1.4715	0.9068	1.4614	0.2684	0.7594	19.0826
ours ($L^*_{D_i}$, $L^{adv*}_{D_i}$)	$\lambda = 0.5, \gamma = 0.5$	1.4787	0.8933	1.3735	0.2702	0.7615	19.17
ours ($L^*_{D_i}$, $L^{adv*}_{D_i}$)	$\lambda = 0.5, \gamma = 5$	1.3997	0.8427	0.8844	0.2508	0.7515	19.2911
ours ($L^{adv*}_{D_i}$)	$\lambda = 0.5, \gamma = 0.1$	1.4701	0.905	1.3959	0.2714	0.7541	19.1149
ours ($L^{adv*}_{D_i}$)	$\lambda = 0.5, \gamma = 0.5$	1.4702	0.8841	1.3156	0.2527	0.7542	19.1338
ours ($L^{adv*}_{D_i}$)	$\lambda = 0.5, \gamma = 5$	1.4239	0.856	0.956	0.2198	0.7543	19.2189

4.3. Ablation Studies

In order to illustrate the effect of the gamma on each metric for both type of losses, we perform extensive experiments and the results are summarized in the Figure 4. Gamma indicates different weight for adversarial loss of D_i, see Equation (6). We alter gamma as we intend to see the effect of infrared images. For our experiments, we choose γ = 0.1, 0.2, 0.3, 0.5, 0.8, 2, 3 and 5. We can observe that out of the six metrics, four gives highest results for gamma equal to 0.1 and 0.2. However, for each metric except SCD the values are highest between 0 and 1. After 1, the values start to decrease.

We show the performance results with two techniques of using the new loss function. We also perform experiments with different values of gamma. We witness that the γ between 0 and 1 yields the best results for each quality metrics expect SCD in Figure 4. We also observe the values for both techniques of using the new loss in D_i only, and in both D_i and G and see that the loss used in both D_i and G delivers good performance overall.

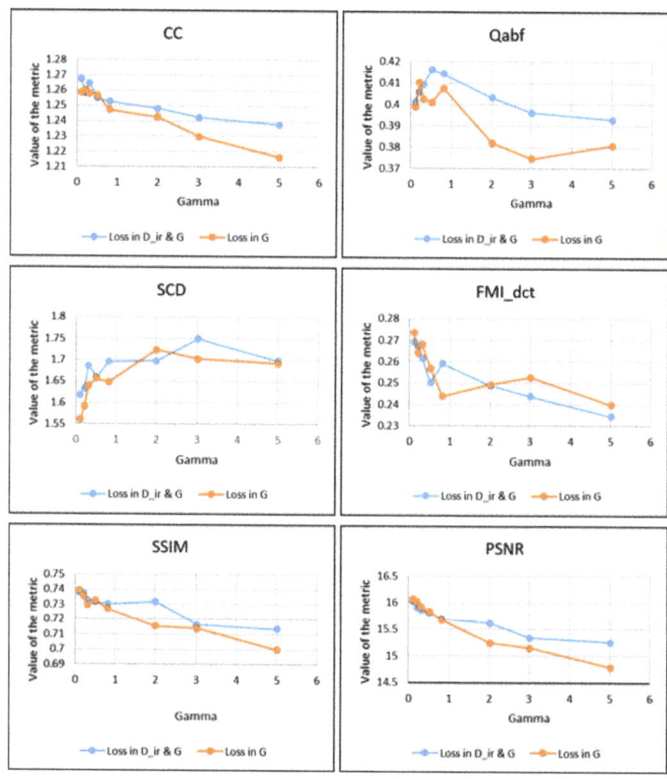

Figure 4. Values of each metric for images from RoadScene Dataset with different γ. Lambda is fixed to 0.5 for this analysis. Type of the loss used is shown in the legends.

There are two ways to construct the weight maps equal to the size of the output of discriminator, average pooling and maximum pooling. In Table 5, we analyze the effect of using both pooling strategies for the construction of the weights.The settings for this analysis include the weighted loss used in both D_i and G with $\lambda = 0.5$ and $\gamma = 0.1$. Based on these results, it is quite evident that average pooling performs better than maximum pooling.

Table 5. The average values of quality metrics for 20 fused images on two datasets with different pooling methods. The best values are indicated in blue.

Dataset	Methods	CC	$MSSSIM$	SCD [33]	FMI_{dct} [34]	$SSIM_a$	PSNR
RoadScene	Avg-Pooling	1.2675	0.8655	1.6100	0.2691	0.7381	16.0222
RoadScene	Max-Pooling	1.2637	0.8617	1.5528	0.2671	0.7375	16.0355
Private	Avg-Pooling	1.4715	0.9068	1.4614	0.2684	0.7594	19.0826
Private	Max-Pooling	1.4467	0.8988	1.3488	0.2594	0.7508	19.224

5. Conclusions

This paper presents a new end-to-end trainable framework where patchGAN is used with dual discriminators. The advantage of using the patchGAN is that it tries to signify if each $N \times N$ patch in an image is real or fake rather than determining the entire image, allowing us to observe features that are otherwise hard to perceive. The fundamental characteristic of the proposed method is the weights derived from the infrared images. These weights are utilized in defining the new loss function which can be used in two ways; (1) Use the new weighted loss in infrared discriminator and as an adversarial loss in generator, (2) Use the new loss as adversarial loss in generator only. These weight maps can help achieve our objective of accurately extracting the information from the highly informative regions of infrared as well as visible images. The experiments are conducted on RoadScene and our private dataset to evaluate the performance of our proposed method qualitatively as well as quantitatively. The experimental results indicate that the images fused by the proposed method contain more details and are more vivid. The proposed technique is simple yet effective and achieves better results than the state-of-the-art methods.

Author Contributions: Conceptualization, J.-H.K. and Y.H.; methodology, S.M. and J.-H.K.; software, S.M.; validation, S.M. and J.-H.K.; writing—original draft preparation, S.M.; writing—review and editing, Y.H.; visualization, S.M.; supervision, J.-H.K. and Y.H.; project administration, Y.H.; funding acquisition, Y.H. All authors have read and agreed to the published version of the manuscript.

Funding: This work was partially supported by the Grand Information Technology Research support program (IITP-2020-0-01462), and by the National Research Foundation of Korea (NRF) grant funded by the Korea government (MSIT) (No. 2020R1F1A1077110).

Conflicts of Interest: The authors declare no conflict of interest.

References

1. Liu, Y.; Dong, L.; Ji, Y.; Xu, W. Infrared and Visible Image Fusion through Details Preservation. *Sensors* **2019**, *19*, 4556. [CrossRef] [PubMed]
2. Liu, Y.; Yang, X.; Zhang, R.; Albertini, M.K.; Celik, T.; Jeon, G. Entropy-Based Image Fusion with Joint Sparse Representation and Rolling Guidance Filter. *Entropy* **2020**, *22*, 118. [CrossRef]
3. Jiang, W.; Yang, X.; Wu, W.; Liu, K.; Ahmad, A.; Sangaiah, A.K.; Jeon, G. Medical images fusion by using weighted least squares filter and sparse representation. *Comput. Electr. Eng.* **2018**, *67*, 252–266. [CrossRef]
4. Shao, Z.; Wu, W.; Guo, S. IHS-GTF: A Fusion Method for Optical and Synthetic Aperture Radar Data. *Remote Sens.* **2020**, *12*, 2796. [CrossRef]
5. Chipman, L.; Orr, T.; Graham, L. Wavelets and image fusion. In Proceedings of the International Conference on Image Processing, Washington, DC, USA, 23–26 October 1995; Volume 3, pp. 248–251.
6. Lewis, J.; O'Callaghan, R.; Nikolov, S.; Bull, D.; Canagarajah, N. Pixel- and region-based image fusion with complex wavelets. *Inf. Fusion* **2007**, *8*, 119–130. [CrossRef]
7. Xiang, T.Z.; Yan, L.; Gao, R. A fusion algorithm for infrared and visible images based on adaptive dual-channel unit-linking PCNN in NSCT domain. *Infrared Phys. Technol.* **2015**, *69*, 53–61. [CrossRef]
8. Li, H.; Wu, X.J.; Kittler, J. Infrared and visible image fusion using a deep learning framework. In Proceedings of the 2018 24th International Conference on Pattern Recognition (ICPR), Beijing, China, 20–24 August 2018; pp. 2705–2710.
9. Li, H.; Wu, X.J. DenseFuse: A fusion approach to infrared and visible images. *IEEE Trans. Image Process.* **2018**, *28*, 2614–2623. [CrossRef] [PubMed]
10. Ma, J.; Ma, Y.; Li, C. Infrared and visible image fusion methods and applications: A survey. *Inf. Fusion* **2019**, *45*, 153–178. [CrossRef]
11. Xu, D.; Wang, Y.; Xu, S.; Zhu, K.; Zhang, N.; Zhang, X. Infrared and Visible Image Fusion with a Generative Adversarial Network and a Residual Network. *Appl. Sci.* **2020**, *10*, 554. [CrossRef]
12. Ma, J.; Xu, H.; Jiang, J.; Mei, X.; Zhang, X.P. DDcGAN: A Dual-Discriminator Conditional Generative Adversarial Network for Multi-Resolution Image Fusion. *IEEE Trans. Image Process.* **2020**, *29*, 4980–4995. [CrossRef]
13. Zhao, F.; Zhao, W.; Yao, L.; Liu, Y. Self-supervised feature adaption for infrared and visible image fusion. *Inf. Fusion* **2021**, *76*, 189–203. [CrossRef]
14. Li, S.; Kang, X.; Hu, J. Image Fusion With Guided Filtering. *IEEE Trans. Image Process.* **2013**, *22*, 2864–2875. [PubMed]
15. Prabhakar, K.; Srikar, V.; Babu, R. DeepFuse: A Deep Unsupervised Approach for Exposure Fusion with Extreme Exposure Image Pairs. In Proceedings of the 2017 IEEE International Conference on Computer Vision (ICCV), Venice, Italy, 22–29 October 2017; pp. 4724–4732.
16. Sun, C.; Zhang, C.; Xiong, N. Infrared and Visible Image Fusion Techniques Based on Deep Learning: A Review. *Electronics* **2020**, *9*, 2162. [CrossRef]
17. Isola, P.; Zhu, J.Y.; Zhou, T.; Efros, A.A. Image-To-Image Translation With Conditional Adversarial Networks. In Proceedings of the IEEE Conference on Computer Vision and Pattern Recognition (CVPR), Honolulu, HI, USA, 21–26 July 2017.
18. Goodfellow, I.; Pouget-Abadie, J.; Mirza, M.; Xu, B.; Warde-Farley, D.; Ozair, S.; Courville, A.; Bengio, Y. Generative adversarial nets. In Proceedings of the Advances in Neural Information Processing Systems 27: 28th Annual Conference on Neural Information Processing Systems 2014, Montreal, QC, Canada, 8–13 December 2014; pp. 2672–2680.
19. Mirza, M.; Osindero, S. Conditional Generative Adversarial Nets. *arXiv* **2014**, arXiv:1411.1784.

20. Mao, X.; Li, Q.; Xie, H.; Lau, R.Y.; Wang, Z.; Paul Smolley, S. Least squares generative adversarial networks. In Proceedings of the IEEE International Conference on Computer Vision, Venice, Italy, 22–29 October 2017; pp. 2794–2802.
21. Arjovsky, M.; Chintala, S.; Bottou, L. Wasserstein generative adversarial networks. In Proceedings of the International Conference on Machine Learning, Sudney, Australia, 7–9 August 2017; pp. 214–223.
22. Ma, J.; Yu, W.; Liang, P.; Li, C.; Jiang, J. FusionGAN: A generative adversarial network for infrared and visible image fusion. *Inf. Fusion* **2019**, *48*, 11–26. [CrossRef]
23. Xu, H.; Ma, J.; Jiang, J.; Guo, X.; Ling, H. U2Fusion: A unified unsupervised image fusion network. *IEEE Trans. Pattern Anal. Mach. Intell.* **2020**. [CrossRef]
24. Li, H.; Wu, X.J.; Durrani, T. NestFuse: An Infrared and Visible Image Fusion Architecture Based on Nest Connection and Spatial/Channel Attention Models. *IEEE Trans. Instrum. Meas.* **2020**, *69*, 9645–9656. [CrossRef]
25. Li, C.; Wand, M. Precomputed real-time texture synthesis with markovian generative adversarial networks. In Proceedings of the European Conference on Computer Vision, Amsterdam, The Netherlands, 8–16 October 2016; pp. 702–716.
26. Demir, U.; Ünal, G.B. Patch-Based Image Inpainting with Generative Adversarial Networks. *arXiv* **2018**, arXiv:1803.07422.
27. Wu, C.; Du, H.; Wu, Q.; Zhang, S. Image Text Deblurring Method Based on Generative Adversarial Network. *Electronics* **2020**, *9*, 220.
28. Liu, X.; Gao, Z.; Chen, B.M. MLFcGAN: Multilevel feature fusion-based conditional GAN for underwater image color correction. *IEEE Geosci. Remote Sens. Lett.* **2019**, *17*, 1488–1492. [CrossRef]
29. Radford, A.; Metz, L.; Chintala, S. Unsupervised Representation Learning with Deep Convolutional Generative Adversarial Networks. *arXiv* **2016**, arXiv:1511.06434.
30. Ioffe, S.; Szegedy, C. Batch Normalization: Accelerating Deep Network Training by Reducing Internal Covariate Shift. In Proceedings of the 32nd International Conference on Machine Learning, Lille, France, 6–11 July 2015; pp. 448–456.
31. Ronneberger, O.; Fischer, P.; Brox, T. U-Net: Convolutional Networks for Biomedical Image Segmentation. In Proceeding of the International Conference on Medical Image Computing and Computer-Assisted Intervention, Munich, Germany, 5–9 October 2015; pp. 234–241.
32. Zuo, Y.; Liu, J.; Bai, G.; Wang, X.; Sun, M. Airborne Infrared and Visible Image Fusion Combined with Region Segmentation. *Sensors* **2017**, *17*, 1127. [CrossRef] [PubMed]
33. Aslantas, V.; Bendes, E. A new image quality metric for image fusion: The sum of the correlations of differences. *AEU-Int. J. Electron. Commun.* **2015**, *69*, 1890–1896. [CrossRef]
34. Haghighat, M.; Razian, M.A. Fast-FMI: Non-reference image fusion metric. In Proceedings of the 2014 IEEE 8th International Conference on Application of Information and Communication Technologies (AICT), Astana, Kazakhstan, 15–17 October 2014; pp. 1–3.

Article

Design and Implementation of Novel Efficient Full Adder/Subtractor Circuits Based on Quantum-Dot Cellular Automata Technology

Mohsen Vahabi [1], Pavel Lyakhov [2,*] and Ali Newaz Bahar [3]

1 Department of Electrical Engineering, Islamic Azad University of Science and Research Tehran (Kerman) Branch, Kerman 7718184483, Iran; mohsen.vahabi@iauk.ac.ir
2 Department of Automation and Control Processes, Saint Petersburg Electrotechnical University "LETI", 197376 Saint Petersburg, Russia
3 Department of Information and Communication Technology (ICT), Mawlana Bhashani Science and Technology University, Tangail 1902, Bangladesh; bahar.ict@mbstu.ac.bd
* Correspondence: ljahov@mail.ru

Abstract: One of the emerging technologies at the nanoscale level is the Quantum-Dot Cellular Automata (QCA) technology, which is a potential alternative to conventional CMOS technology due to its high speed, low power consumption, low latency, and possible implementation at the atomic and molecular levels. Adders are one of the most basic digital computing circuits and one of the main building blocks of VLSI systems, such as various microprocessors and processors. Many research studies have been focusing on computable digital computing circuits. The design of a Full Adder/Subtractor (FA/S), a composite and computing circuit, performing both the addition and the subtraction processes, is of particular importance. This paper implements three new Full Adder/Subtractor circuits with the lowest number of cells, lowest area, lowest latency, and a coplanar (single-layer) circuit design, as was shown by comparing the results obtained with those of the best previous works on this topic.

Keywords: Quantum-Dot Cellular Automata (QCA); Full Adder/Subtractor (FA/S); coplanar

Citation: Vahabi, M.; Lyakhov, P.; Bahar, A.N. Design and Implementation of Novel Efficient Full Adder/Subtractor Circuits Based on Quantum-Dot Cellular Automata Technology. *Appl. Sci.* **2021**, *11*, 8717. https://doi.org/10.3390/app11188717

Academic Editor: Vladimir M. Fomin

Received: 19 July 2021
Accepted: 15 September 2021
Published: 18 September 2021

1. Introduction

The QCA technology, with its unique features such as minimal dimensions, high speed, very low latency, low power consumption, and high operating frequency [1], has attracted the attention of many researchers and scientists as a new method of communication and computation. It has introduced significant novelties in the field of computer science and logic circuits. Adders are one of the most fundamental computational circuits of digital logic and have attracted researchers' attention. Adders are one of the main building blocks of many VLSI systems, such as various microprocessors and processors. Are the new designs aiming at optimizing the relevant blocks compatible with the development of this technology? A complete Adder/Subtractor design with a simple structure and low power consumption can significantly simplify digital circuits. A Full Adder/Subtractor design should include a composite computations circuit and allow performing both addition and subtraction processes. One of the problems in creating hybrid courses is the appropriate composition of wires crossover to reduce costs.

Due to the high price and increasing circuit complexity, a multilayer crossovers design in the implementation of QCA circuits is not desirable (favorable) [2,3]. To achieve coplanar crossovers, it was suggested to rotate the QCA cells, but due to the coexistence of two types of QCA cells, this caused some problems, such as low stability and high implementation cost. Therefore, a design including this type of cells is not desirable [2,4]. The best method for designing QCA circuits is based on the use of 90-degree cells with non-adjacent clock phases (four clock phases) to develop the crossover wires in a single layer [2,5]. Therefore, we used these two types of crossover for this research.

For this reason, The proposed designs, due to their coplanar structure and to the fact that they do not require other layers, have a reduced number of cells, occupied area, and delay. The remainder of this article is organized as follows. Section 2 (Background) provides an overview of QCA and previous literature. Section 3 (Proposed Circuits) presents the proposed architecture of Full Adder/Subtractor circuits. In Section 4 (Guidelines Performance Evaluation), we compare the proposed designs with previous architectures. In Section 5 (Conclusion), we discuss our conclusions.

2. Background

2.1. The Basis of Quantum-Dot Cellular Automata (QCA) Technology

This technology is based on QCA cells, and the basis of the QCA cell can represent a logical bit with occupied space in the nanoscale. A QCA cell includes two electrons, and, based on the Coulombic repulsion created between two electrons, two logical values of "0" and "1" are possible. The QCA cells are square, as shown in Figure 1. Each enclosure consists of four holes. The two electrons are trapped inside and can move freely between the holes; by placing two electrons in four spots, six different states are created, which is impossible due to Coulombic repulsion forces between the electrons. As a result, to satisfy these forces, electrons are placed inside the holes with as far apart as possible, until the Coulombic repulsion law is satisfied. Depending on the location of the electrons and their diameter, two structures are created; by the establishment of two electrons in each of these two poles, two different states are created. With these two types of systems, two logical values can be obtained; we will consider one of the logical values. We attribute these two polar structures 1 and –1 to the logical values of "1" and "0", respectively; the poles at 1 and –1 same are those of the square cells, as shown in Figure 1 [6–8].

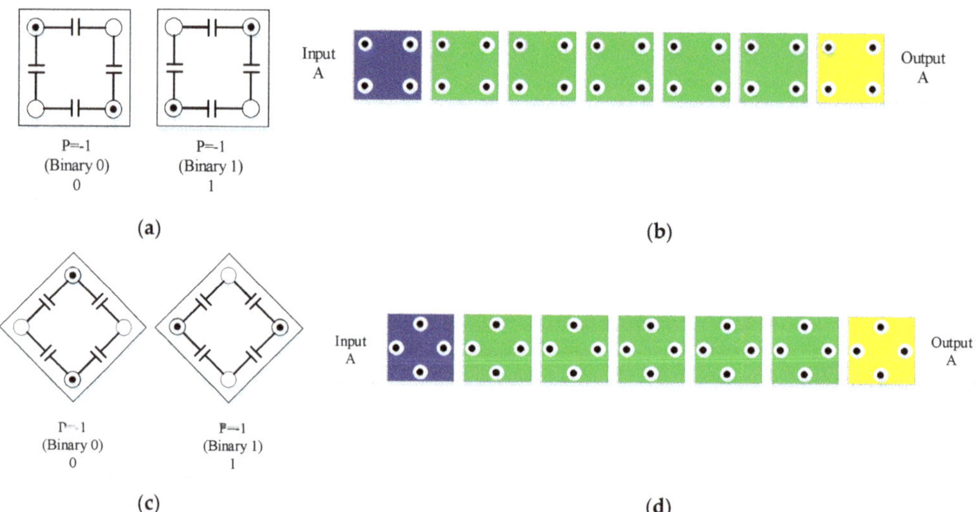

Figure 1. (**a**) Normal QCA cells' structure, (**b**) Normal QCA wire's structure, (**c**) Rotated QCA cells' structure and (**d**) Rotated QCA wire's structure.

When the electrons move inside the cell, they tunnel between the holes. Then, the moving of the electrons inside the cell is similar to a nonlinear move, and the Coulombic repulsion force is not exerted just between the electrons inside a cell. However, as shown in Figure 2, each cell adjacent to this one, which has a logical value, is affected and affects the next adjacent cell that has no value, converting it to its value [6–8].

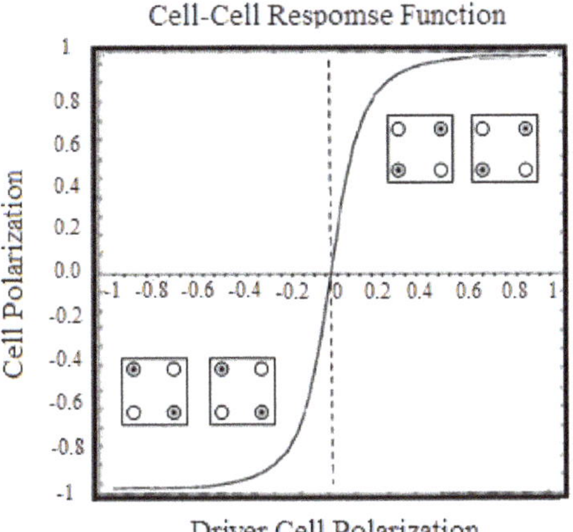

Figure 2. Behavior of adjacent cells.

2.2. QCA Four-Phase Clock

A four-clocking phase scheme for the QCA is shown in Figure 3. As shown in the figure, the barriers (potential barriers) rise during the first clock phase (switch). At the beginning of this phase, the borders are low, and the QCA cell is unpolarized; in this state, under the effect Columbic repulsion, the cell receives data from its adjacent cells. Then, with the barriers rising, the QCA cells are polarized according to their input drive modes, and at the end of this clock phase, the borders are high enough to prevent electron tunneling. As a result, the cell is locked. It is in this phase that the actual switching happens. During the second phase of the clock (hold), the barriers remain high. In this phase, the cell is relatively stable and transmits its data to the adjacent cells. The walls gradually decrease during the third clock phase (release), and the cell becomes unstable. In this phase, the cell is allowed to lose its polarization (unpolarized). During the fourth clock phase (relax phase), the cell barriers are in the lowest state, and the cells remain unpolarized. The cell is not used in this phase. After the end of this phase, the cell enters the switch phase again [3,9].

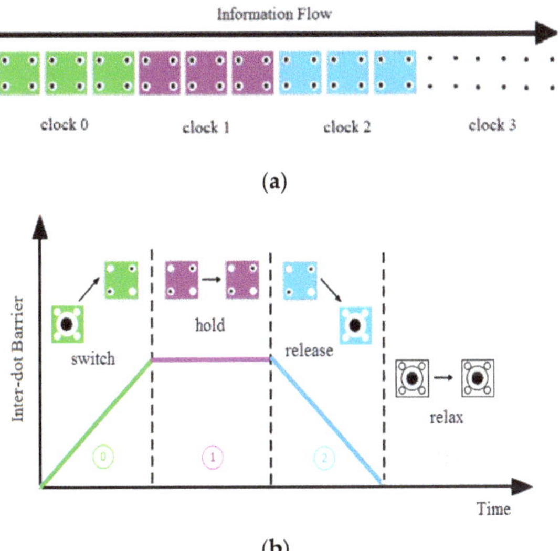

Figure 3. (**a**) Clock phases and (**b**) QCA four-phase clock mechanism.

2.3. QCA Four-Phase Clock

One of the gates used in logic circuits is the inverter gate (Not gate). A type of inverter gate used in QCA technology is shown in Figure 4. It is used for inverting the desired signal as required [10,11].

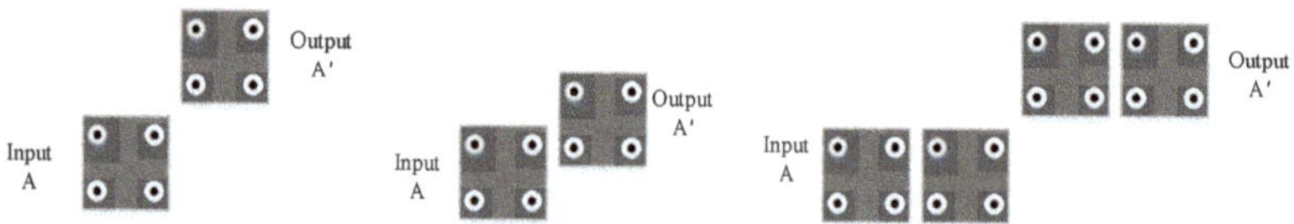

Figure 4. Two different Not-gate in QCA based on 90° cells.

One of the most usable logic gates in QCA technology is the majority gate. This gate has an odd number of inputs and one output. In other words, the output cell value (output cell polarization) is determined according to the logical value of the majority inputs. As a result, the output cell value is determined based on the majority of inputs [8,10]. Figure 5a shows an example of this gate.

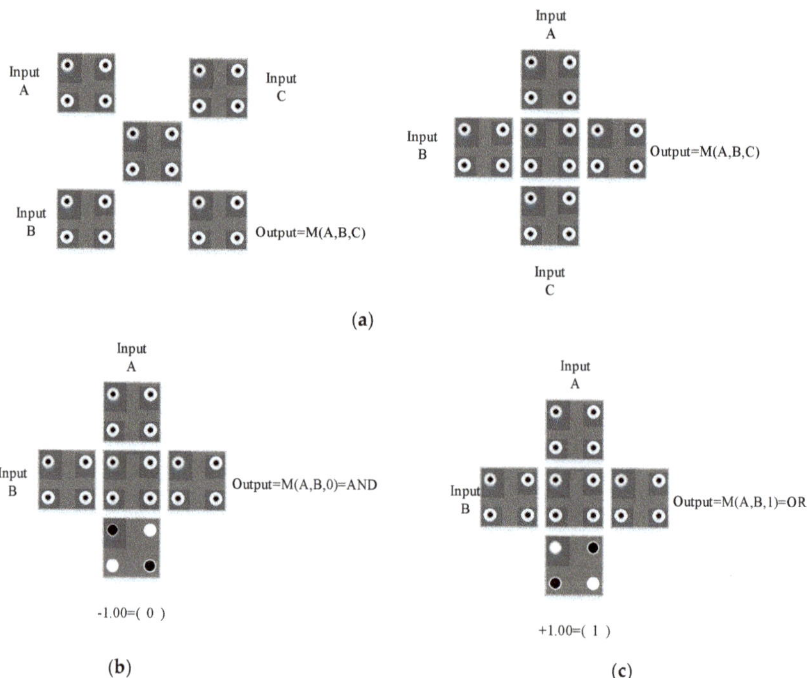

Figure 5. (**a**) QCA with implementation of the majority gate, (**b**) QCA with implementation of the AND Gate with two inputs and (**c**) QCA with implementation of the OR Gate with two inputs.

By stabilizing (fixed) one of the inputs of the majority gate and considering a logical value "0" (polarization −1), the AND gate is generated [10,12]. Figure 5b shows a two-input AND gate.

The OR gate is generated by fixing one of the majority gate inputs and considering a logical value "1" (polarization +1) [10,12]. Figure 5c shows a two-input OR gate.

2.4. Related Work

In a previous paper [13], the architecture of a Full Adder/Full Subtractor with a multilayer crossover design was described (Figure 6). This type of multilayer design requires a larger consumption area than we planned. It has also more cells and delays with respect to our designs, so its cost is very high. As a result, our proposed technique was implemented using the coplanar method. It has significant advantages over this type of architecture regarding cell number, delay, and area consumption.

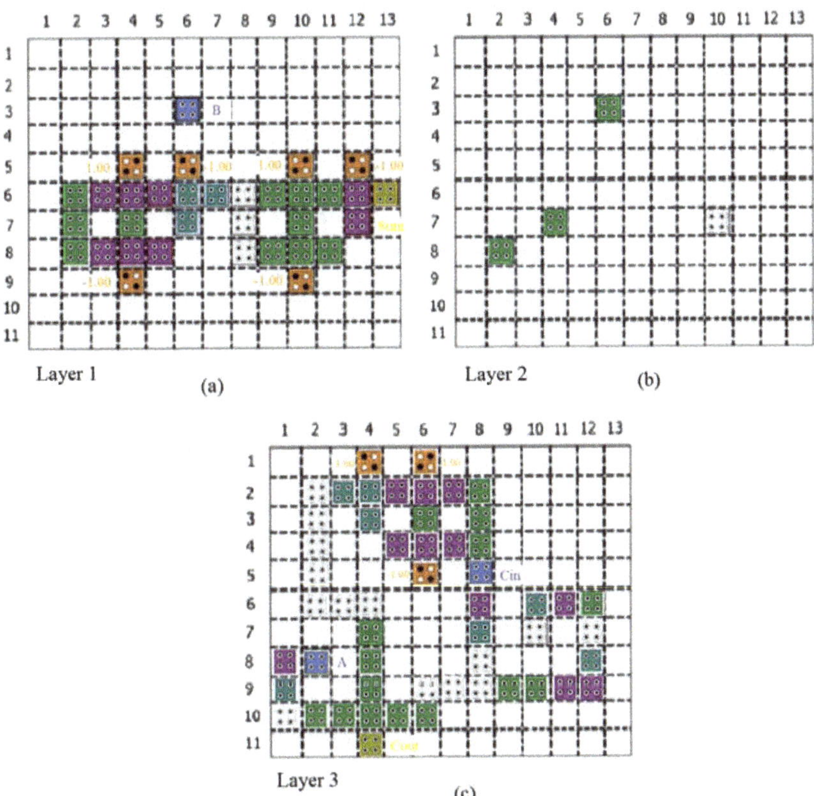

Figure 6. Full Adder/Full Subtractor circuit [13].

In another paper [14], the architecture of a Full Adder/Full Subtractor was also presented (Figure 7). This type of design, due to the high delay and unsuitable carry output, requires another crossover. That leads to an increase in the delay and number of cells. As a result, this architecture is also not suitable. Our proposed design has significant advantages relative to this design [14], such as the number of cells, delay, area of consumption, and therefore cost function.

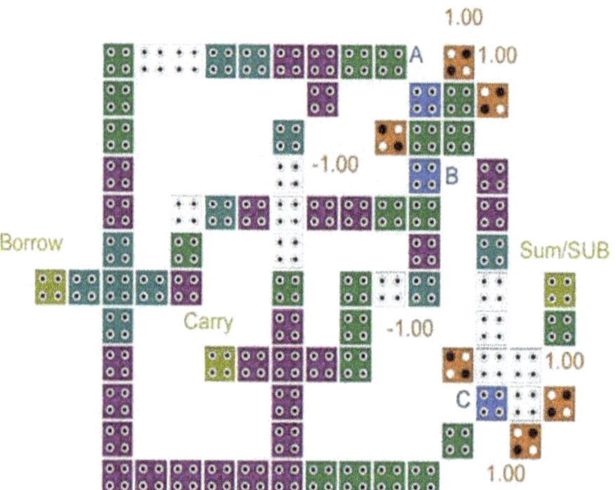

Figure 7. Full Adder/Full Subtractor circuit [14].

Another paper [15] described a Full Adder/Full Subtractor architecture using the coplanar method design (Figure 8). This type of design is not very favorable, because the rotated cells (45° cells) make it more vulnerable and increase the implementation costs. Our method has significant advantages also relative to this design [15], such as the number of cells, delay, and area. Besides, normal cells were used. Our design (C) is 50% better than that design.

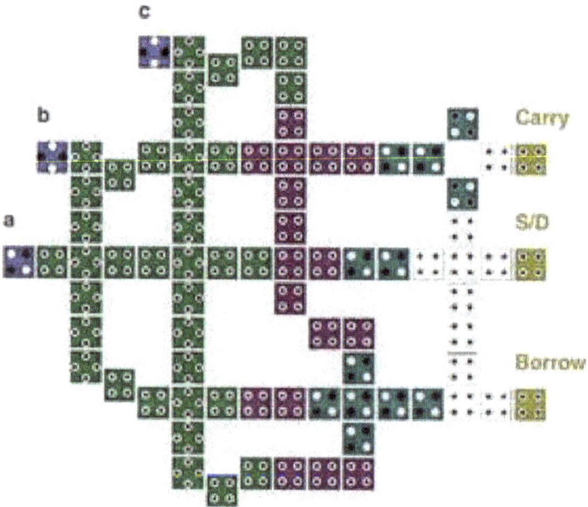

Figure 8. Full Adder/Full Subtractor circuit [15].

In a paper [16], a Full Adder/Full Subtractor circuit architecture, implemented in a coplanar method design was presented. Our proposed designs (A and B) are an addition to the coplanar designs in terms of number cells, area and, therefore, cost of implementation and have significant advantages relative to this design (Figure 9). Despite not using rotated cells in our designs (A and B), their latency is equal to that reported in this previous article. In comparison, our third design delay (C) is 33.34% superior to this design and is ideal in terms of cell number and area.

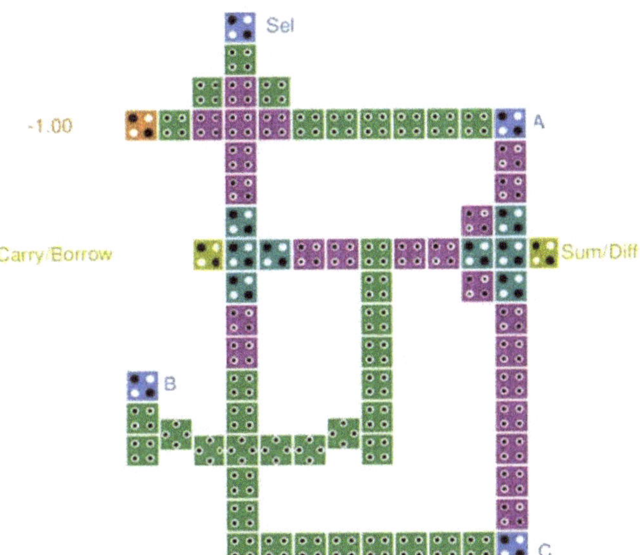

Figure 9. Full Adder/Full Subtractor circuit [16].

Another paper [17] also presented two Full Adder/Full Subtractor circuit designs, as shown in Figures 10 and 11; these designs are also coplanar, but our designs have significant advantages in terms of number of cells, delay, and area.

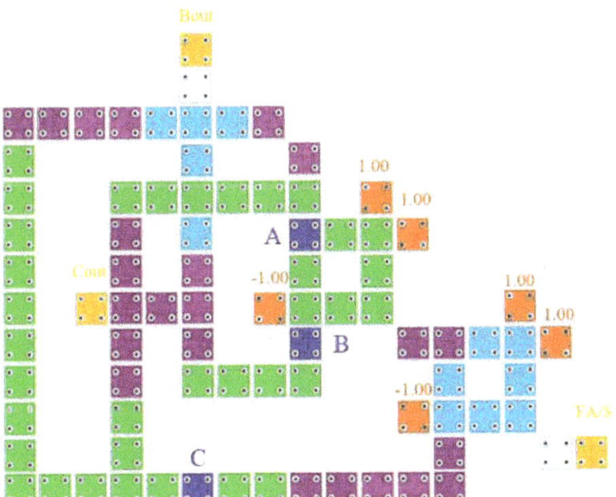

Figure 10. Full Adder/Full Subtractor circuit [17]-a.

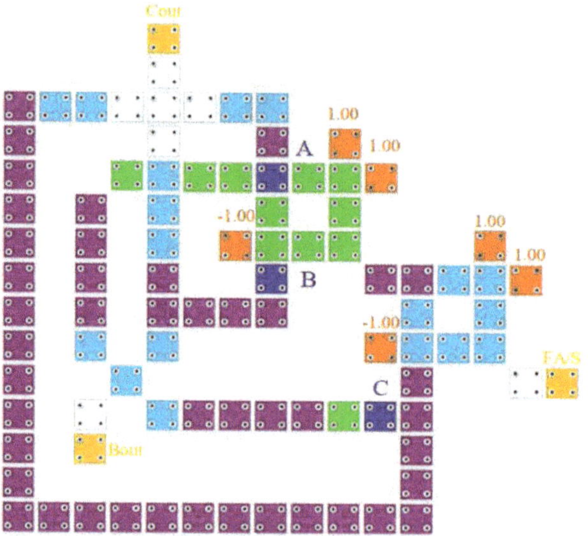

Figure 11. Full Adder/Full Subtractor circuit [17]-b.

3. The Proposed Circuits

In this paper, we designed three new Full Adder/Subtractor (FA/S) circuits based on the XOR gate [18] with the lowest number of cells, smallest consumption circuit area, and lowest latency (delay) relative to the previous best circuits. In cases, a single-layer (coplanar) design was used. Therefore, these circuits are the best examples of design ever made. The two designs (A and B) are coplanar and use only standard cells (90° cells). In the third design (C) which is coplanar, the cells are rotated (45° cells) and also coplanar. The third design shows that the use of this type of cell may reduce the delay of circuit outputs, but, it also reduces the stability and resistance of the circuit compared to the circuits with standard cells.

3.1. FA/S Circuits Design

A Full Adder/Subtractor circuit is a combination circuit where two addition and subtraction operations are performed. This circuit has three inputs (A, B, Cin) and three outputs (S\D, Cout, Bout) [19,20]. Equation (1) is the equation of the output S\D, Equation (2) is the equation of the Cout output, and Equation (3) is the equation of the Bout output. Figure 12, block diagram, and Table 1 show the correct Table of this circuit.

$$S\backslash D = A \oplus B \oplus Cin \quad (1)$$

$$Cout = M(A,B,Cin) = A.B + A.C + B.C \quad (2)$$

$$Bout = M(A',B,Cin) = A'.B + A'.C + B.C \quad (3)$$

This paper designed FA/S circuits with the lowest number of cells, lowest consumable area, and lowest latency (delay), compared to the previous best examples. We used a single-layer (coplanar) design to obtain the best designs ever made. The design is better than previous ones not only in terms of cell number, area, and delay but also because it is based on a single layer. Figure 13 presents a block diagram of these circuits and Figures 14–16, show the implementation of the Proposed Full Adder/Subtractor (FA/S) circuits designs.

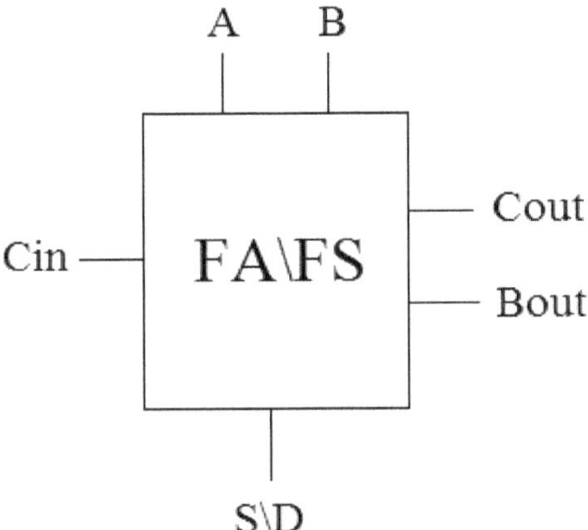

Figure 12. Block diagram of the Full Adder/Subtractor circuit.

Table 1. Truth table of the Full Adder/Subtractor.

Bout	Cout	S\D	Cin	B	A
0	0	0	0	0	0
1	0	1	1	0	0
1	0	1	0	1	0
1	1	0	1	1	0
0	0	1	0	0	1
0	1	0	1	0	1
0	1	0	0	1	1
1	1	1	1	1	1

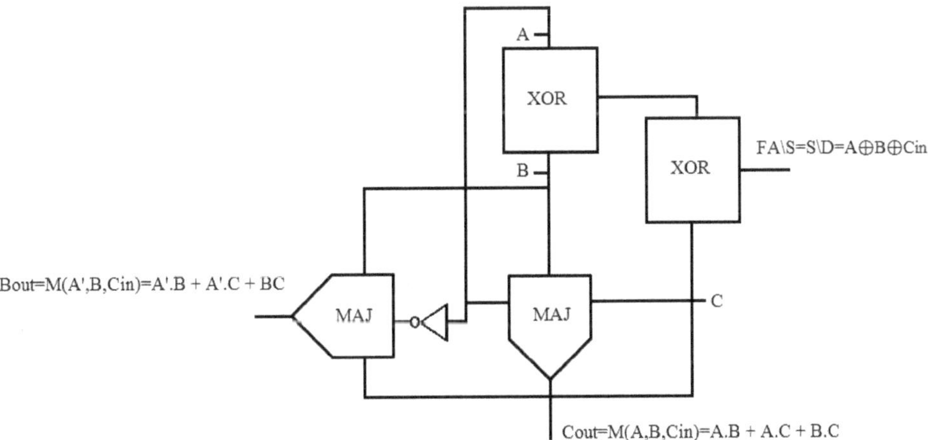

Figure 13. Simulation results of the proposed XOR-gate.

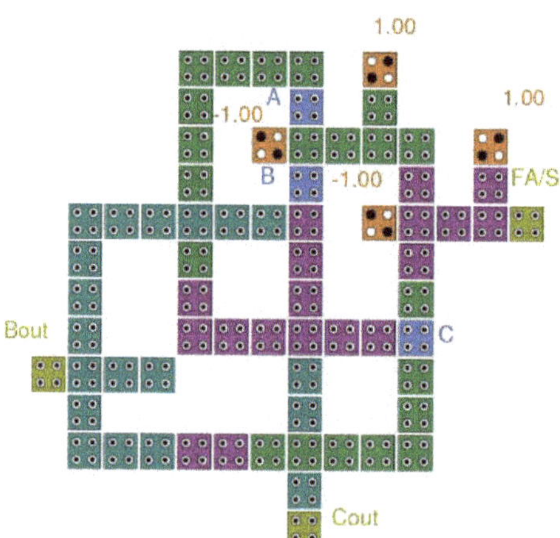

Figure 14. The proposed (A) Full Adder/Full Subtractor circuit.

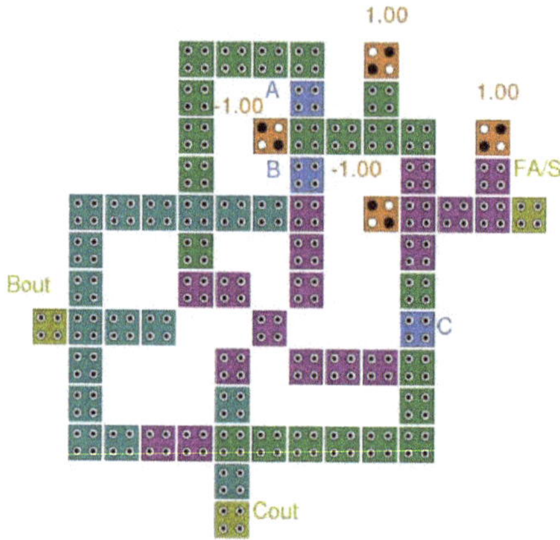

Figure 15. The proposed (B) Full Adder/Full Subtractor circuit.

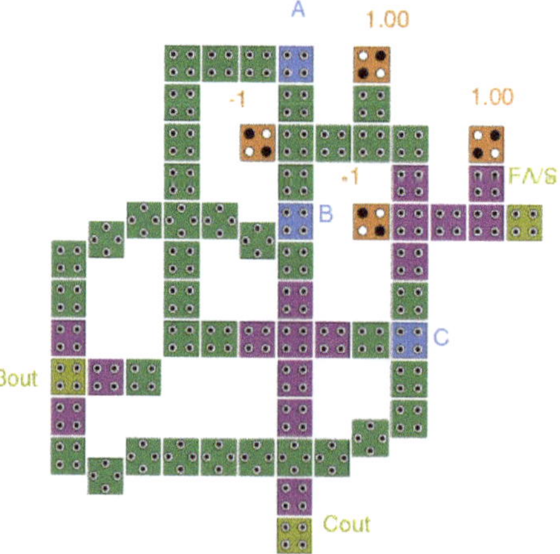

Figure 16. The proposed (C) Full Adder/Full Subtractor circuit.

3.2. Simulation Results

In this section, the simulator outputs of the proposed circuits are shown in Figures 17–19. The output latency of both offered courses (A and B) is the same, and they provide the same simulation outputs. As can be seen, in both circuits, the delay is one clock (four phases). The delay of the third circuit (C) is 0.5 clock (two stages). The proposed Full Adder/Subtractor hybrid circuits combine two addition and subtraction circuits and allow the concurrent performance of both operations.

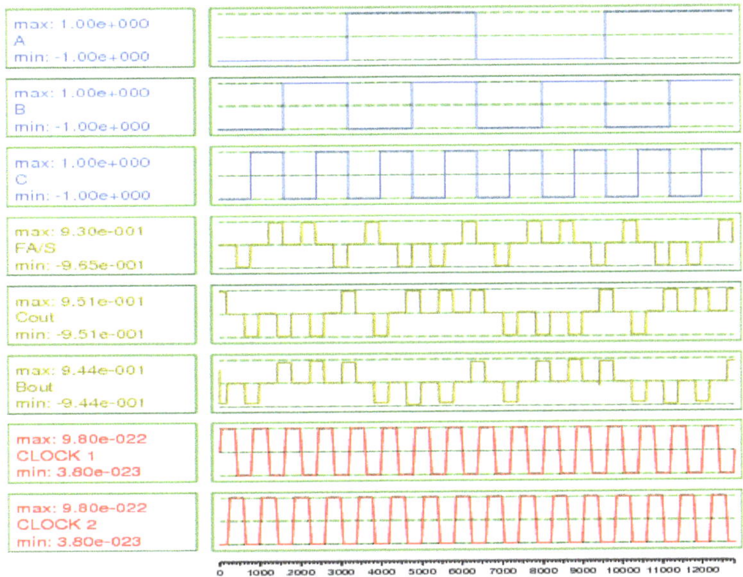

Figure 17. Simulation results for the proposed (A) Full Adder/Full Subtractor circuit.

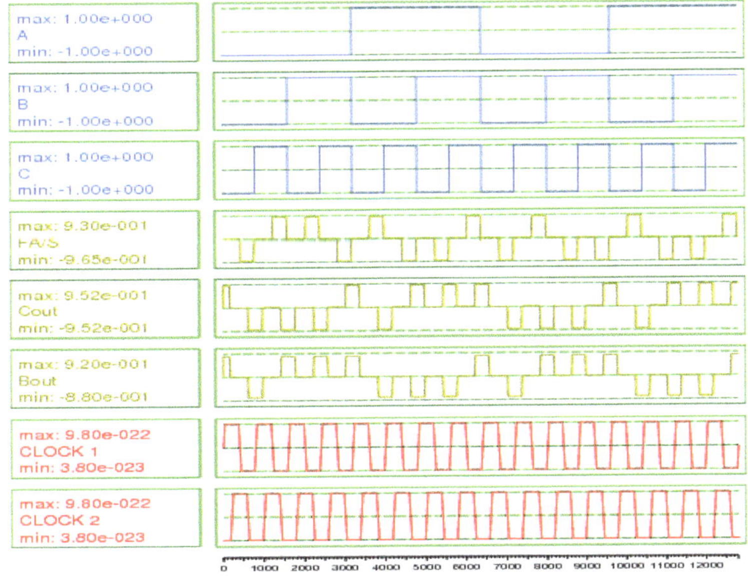

Figure 18. Simulation results for the proposed (B) Full Adder/Full Subtractor circuit.

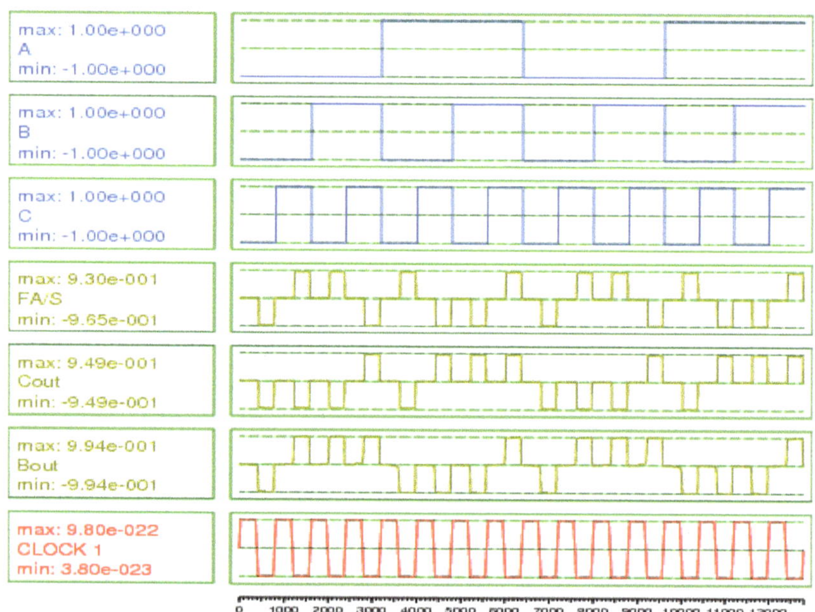

Figure 19. Simulation results for the proposed (C) Full Adder/Full Subtractor circuit.

4. Guidelines of Performance Evaluation

The QCA Designer provided the simulation results. The simulation parameters are presented in Table 2. The proposed design was compared with designs described in previous works. For all circuits designed, parameters including area, delay, and cell numbers are provided. The type of crossover is also presented for a better and more accurate comparison.

Table 2. Simulation parameters for the QCA Designer.

Parameter	Value
Cell width	18 nm
Cell height	18 nm
Dot diameter	5 nm
Number of samples	12,800
Convergence tolerance	0.001
Radius of effect	65 nm
Relative permittivity	12.9
Clock high	9.8×10^{-22} J
Clock low	3.8×10^{-23} J
Clock amplitude factor	2
Layer separation	11.5 nm
Maximum iteration per sample	100

The simulation results are given in Table 3. As can be seen, the proposed circuits were compared with the best circuits previously described. In Table 3, consumption area, delay, and cell number of the proposed Full Adder/Subtractor circuits are compared to those of previous designs.

Table 3. Comparing the Full Adder/Subtractor (FA/S) of this study with those of previous works.

Crossover Type	Latency (clock)	Cell Count	Area (μm^2)	Circuit
Multi-Layer	1.5	90	0.6	[13]
Coplanar (clocking based)	1.5	83	0.09	[14]
Coplanar (rotated cells)	1	82	0.11	[15]
Coplanar (rotated cells)	0.75	75	0.09	[16]
Coplanar (clocking based)	1	92	0.09	[17]-a
Coplanar (clocking based)	1	84	0.09	[17]-b
Coplanar (clocking based)	0.75	68	0.072	Proposed A
Coplanar (clocking based)	0.75	67	0.072	Proposed B
Coplanar (rotated cells)	0.5	65	0.067	Proposed C

As shown in Table 3, our designs (A) and (B) allow reducing the area and power consumption up to 39.1% with respect to previous circuits described in [14,17]. As can be seen, the delay in the proposed designs improved significantly with respect to previous works. Our designs (A) and (B) reduce the delay by 50% in comparison to the designs in [13,14] and by 30% with respect to those in [15,17]. The reduction in the proposed design C, relative to the designs in [13,14], corresponds to 66.66%, whereas it corresponds to 50% in comparison to those in [15], [17]-a and [17]-b, and to 33.33% in comparison to that in [16]. As can be seen, the proposed designs have also the lowest cell number with respect to the other designs. Improvement in the cell number of the proposed design A relative to the designs in [13–17]-a and [17]-b is about 24.45%, 18.07%, 17.07%, 9.34%, 26.09%, and 19.05% respectively; the cell number improvement in the proposed design B relative to [13–17]-a and [17]-b, respectively, is about 25.56%, 19.28%, 18.29%, 10.67%, 27.17%, and 20.24%; finally, the cell number improvement in the proposed design C relative to [13–17]-a and [17]-b, respectively, is about 27.78%, 21.69%, 20.73%, 13.34%, 29.35%, and 22.62%.

5. Conclusions

The FA/S designs using the QCA technology use at least three layers for the crossover, while several techniques use 45° cells. Indeed, only non-adjacent clock phases (four clock phases) are required to design the crossover in a single layer, which is robust and better. However, the coplanar crossover's design using rotated cells can reduce delay in the circuit. In some cases, depending on the type of usage, these two types of design can be used. The circuits' designs proposed in this study are better and preferable than previous designs in terms of number of cells consumed, circuit area, delay, and cost. As a result, these three proposed designs can be used in more extensive and more complex circuits.

Author Contributions: Conceptualization, M.V.; methodology, M.V.; validation, M.V., P.L. and A.N.B.; formal analysis, M.V.; investigation, M.V., P.L. and A.N.B.; writing—original draft preparation, M.V.; writing—review and editing, P.L. and A.N.B.; supervision, P.L. and A.N.B. All authors have read and agreed to the published version of the manuscript.

Funding: This research received no external funding.

Data Availability Statement: Data are contained within the article.

Conflicts of Interest: The authors declare no conflict of interest.

References

1. Mohammadi, Z.; Mohammadi, M. Implementing a one-bit reversible full adder using quantum-dot cellular automata. *Quantum Inf. Process.* **2014**, *13*, 2127–2147. [CrossRef]
2. Tougaw, P.D.; Lent, C.S. Logical devices implemented using quantum cellular automata. *J. Appl. Phys.* **1994**, *75*, 1818–1825. [CrossRef]
3. Cho, H.; Swartzlander, E.E. Adder Designs and Analyses for Quantum-Dot Cellular Automata. *IEEE Trans. Nanotechnol.* **2007**, *6*, 374–383. [CrossRef]

4. Huang, J.; Momenzadeh, M.; Tahoori, M.B.; Lombardi, F. Design and characterization of an and-or-inverter (AOI) gate for QCA implementation. In Proceedings of the 14th ACM Great Lakes Symposium on VLSI 2004, Boston, MA, USA, 26–28 April 2004; pp. 426–429. [CrossRef]
5. Abedi, D.; Jaberipur, G.; Sangsefidi, M. Coplanar full adder in quantum-dot cellular automata via clock-zone-based crossover. *IEEE Trans. Nanotechnol.* **2015**, *14*, 497–504. [CrossRef]
6. Fortes, J.A. Future challenges in VLSI system design. In Proceedings of the IEEE Computer Society Annual Symposium on VLSI, Tampa, FL, USA, 20–21 February 2003; pp. 5–7.
7. Frost, S.E.; Rodrigues, A.F.; Janiszewski, A.W.; Rausch, R.T.; Kogge, P.M. Memory in motion: A study of storage structures in QCA. In Proceedings of the First Workshop on Non-Silicon Computing, Cambridge, MA, USA, 3 February 2002; Volume 2.
8. Wang, W.; Walus, K.; Jullien, G.A. Quantum-dot cellular automata adders. In Proceedings of the 2003 Third IEEE Conference on Nanotechnology, San Francisco, CA, USA, 12–14 August 2003; Volume 1, pp. 461–464.
9. Kianpour, M.; Sabbaghi-Nadooshan, R. Optimized Design of Multiplexor by Quantum-dot Cellular Automata. *Int. J. Nanosci. Nanotechnol.* **2013**, *9*, 15–24.
10. Beigh, M.R.; Mustafa, M.; Ahmad, F. Performance Evaluation of Efficient XOR Structures in Quantum-Dot Cellular Automata (QCA). *Circuits Syst.* **2013**, *4*, 147–156. [CrossRef]
11. Swartzlander, E.E.; Cho, H.; Kong, I.; Kim, S.W. Computer arithmetic implemented with QCA: A progress report. In Proceedings of the 2010 Conference Record of the Forty Fourth Asilomar Conference on Signals, Systems and Computers, Pacific Grove, CA, USA, 7–10 November 2010; pp. 1392–1398.
12. Modi, S.; Tomar, A.S. Logic gate implementations for quantum dot cellular automata. In Proceedings of the 2010 International Conference on Computational Intelligence and Communication Networks, Bhopal, India, 26–28 November 2010.
13. Barughi, Y.Z.; Heikalabad, S.R. A Three-Layer Full Adder/Subtractor Structure in Quantum-Dot Cellular Automata. *Int. J. Theor. Phys.* **2017**, *56*, 2848–2858. [CrossRef]
14. Zoka, S.; Gholami, M. A novel efficient full adder–subtractor in QCA nanotechnology. *Int. Nano Lett.* **2018**, *9*, 51–54. [CrossRef]
15. Sadeghi, M.; Navi, K.; Dolatshahi, M. Novel efficient full adder and full subtractor designs in quantum cellular automata. *J. Supercomput.* **2019**, *76*, 2191–2205. [CrossRef]
16. Raj, M.; Gopalakrishnan, L.; Ko, S.-B. Design and analysis of novel QCA full adder-subtractor. *Int. J. Electron. Lett.* **2020**, *9*, 1–14. [CrossRef]
17. Vahabi, M.; Sabbagh Molahosseini, A. A New Coplanar Full Adder/Subtractor in Quantum-Dot Cellular Automata Technology. *Majlesi J. Telecommun. Devices* **2018**, *7*, 53–63.
18. Chabi, A.M.; Roohi, A.; Khademolhosseini, H.; Sheikhfaal, S.; Angizi, S.; Navi, K.; DeMara, R.F. Towards ultra-efficient QCA reversible circuits. *Microprocess. Microsyst.* **2017**, *49*, 127–138. [CrossRef]
19. Gladshtein, M. Design and simulation of novel adder/subtractors on quantum-dot cellular automata: Radical departure from Boolean logic circuits. *Microelectron. J.* **2013**, *44*, 545–552. [CrossRef]
20. Ahmad, F.; Ahmed, S.; Kakkar, V.; Bhat, G.M.; Bahar, A.N.; Wani, S. Modular design of ultra-efficient reversible full adder-subtractor in QCA with power dissipation analysis. *Int. J. Theor. Phys.* **2018**, *57*, 2863–2880. [CrossRef]

Article

An Algorithm for Fast Multiplication of Kaluza Numbers

Aleksandr Cariow †, Galina Cariowa † and Janusz P. Paplinski *,†

Faculty of Computer Science and Information Technology, West Pomeranian University of Technology, Szczecin, Żołnierska 49, 71-210 Szczecin, Poland; acariow@wi.zut.edu.pl (A.C.); gcariowa@wi.zut.edu.pl (G.C.)

* Correspondence: janusz.paplinski@zut.edu.pl

† These authors contributed equally to this work.

Abstract: This paper presents a new algorithm for multiplying two Kaluza numbers. Performing this operation directly requires 1024 real multiplications and 992 real additions. We presented in a previous paper an effective algorithm that can compute the same result with only 512 real multiplications and 576 real additions. More effective solutions have not yet been proposed. Nevertheless, it turned out that an even more interesting solution could be found that would further reduce the computational complexity of this operation. In this article, we propose a new algorithm that allows one to calculate the product of two Kaluza numbers using only 192 multiplications and 384 additions of real numbers.

Keywords: convolutional neural networks; fast algorithms; hypercomplex number multiplication; Kaluza numbers

Citation: Cariow, A.; Cariowa, G.; Paplinski, J.P. An Algorithm for Fast Multiplication of Kaluza Numbers. *Appl. Sci.* **2021**, *11*, 8203. https://doi.org/10.3390/app11178203

Academic Editor: Pavel Lyakhov

Received: 6 August 2021
Accepted: 1 September 2021
Published: 3 September 2021

1. Introduction

The permanent development of the theory and practice of data processing, as well as the need to solve increasingly complex problems of computational intelligence, inspire the use of complex and advanced mathematical methods and formalisms to represent and process big multidimensional data arrays. A convenient formalism for representing big data arrays is the high-dimensional number system. For a long time, high-dimensional number systems have been used in physics and mathematics for modeling complex systems and physical phenomena. Today, hypercomplex numbers [1] are also used in various fields of data processing, including digital signal and image processing, machine graphics, telecommunications, and cryptography [2–10]. However, their use in brain-inspired computation and neural networks has been largely limited due to the lack of comprehensive and all-inclusive information processing and deep learning techniques. Although there has been a number of research articles addressing the use of quaternions and octonions, higher-dimensional numbers remain a largely open problem [11–22]. Recently, new articles appeared in open access that presented a sedenion-based neural network [23,24]. The expediency of using numerical systems of higher dimensions was also noted. Thus, the object of our research was hypercomplex-valued convolutional neural networks using 32-dimensional Kaluza numbers.

In advanced hypercomplex-valued convolutional neural networks, multiplying hypercomplex numbers is the most time-consuming arithmetic operation. The reason for this is that the addition of N-dimensional hypercomplex numbers requires N real additions, while the multiplication of these numbers already requires $N(N-1)$ real additions and N^2 real multiplication. It is easy to see that the increasing of dimensions of hypercomplex numbers increases the computational complexity of the multiplication. Therefore, reducing the computational complexity of the multiplication of hypercomplex numbers is an important scientific and engineering problem. The original algorithm for computing the product of Kaluza numbers was described in [25], but we found a more efficient solution. The purpose of this article is to present our new solution.

2. Preliminary Remarks

In all likelihood, the rules for constructing Kaluza numbers were first described in [26]. In article [25], based on these rules, a multiplication table for the imaginary units of the Kaluza number was constructed. A Kaluza number is defined as follows:

$$d = d_0 + \sum_{n=1}^{31} d_n e_n,$$

where $N = 2^{m-1}$ and $\{d_n\}$ for $n = 1, 2, \ldots, 31$ are real numbers, and $\{e_n\}$ for $n = 1, 2, \ldots, 31$ are the imaginary units.

Imaginary units $e_1, e_2, \ldots, e_m$ are called principal, and the remaining imaginary units are expressed through them using the formula:

$$e_s = e_p, e_q, \ldots e_r,$$

where $1 \leq p < q < \cdots < r \leq m$.

All kinds of works of imaginary units are entirely based on established rules:

$$e_p^2 = \epsilon_p; \quad \epsilon_q \epsilon_p = \alpha_{pq} \epsilon_p \epsilon_q; \quad p < q; \quad pq = 1, 2, \ldots, m$$

For Kaluza numbers [26]:

$$m = 5, \quad \epsilon_1 = \epsilon_2 = 1, \quad \epsilon_2 = \epsilon_3 = -1, \quad \alpha_{pq} = -1$$

Using the above rules, the results of all possible products of imaginary units of Kaluza numbers can be summarized in the following tables [25]: Tables 1–4. For conveniens of notation we represents each element e_i in the tables by its subscript i, and we set $i = e_i$.

Table 1. Multiplication rules of Kaluza numbers for $e_0, e_1, \ldots, e_{15}$ and $e_0, e_1, \ldots, e_{15}$ (elements e_i denoted by their subscripts, i.e., $i = e_i$).

ine×	0	1	2	3	4	5	6	7	8	9	10	11	12	13	14	15
*ine*0	0	1	2	3	4	5	6	7	8	9	10	11	12	13	14	15
1	1	0	6	7	8	9	2	3	4	5	16	17	18	19	20	21
2	2	−6	0	10	11	12	−1	−16	−17	−18	3	4	5	22	23	24
3	3	−7	−10	−0	13	14	16	1	−19	−20	2	−22	−23	−4	−5	25
4	4	−8	−11	−13	−0	15	17	19	1	−21	22	2	−24	3	−25	−5
5	5	−9	−12	−14	−15	−0	18	20	21	1	23	24	2	25	3	4
6	6	−2	1	16	17	18	−0	−10	−11	−12	7	8	9	26	27	28
7	7	−3	−16	−1	19	20	10	0	−13	−14	6	−26	−27	−8	−9	29
8	8	−4	−17	−19	−1	21	11	13	0	−15	26	6	−28	7	−29	−9
9	9	−5	−18	−20	−21	−1	12	14	15	0	27	28	6	29	7	8
10	10	16	−3	−2	22	23	−7	−6	26	27	0	−13	−14	−11	−12	30
11	11	17	−4	−22	−2	24	−8	−26	−6	28	13	0	−15	10	−30	−12
12	12	18	−5	−23	−24	−2	−9	−27	−28	−6	14	15	0	30	10	11
13	13	19	22	4	−3	25	26	8	−7	29	11	−10	30	−0	15	−14
14	14	20	23	5	−25	−3	27	9	−29	−7	12	−30	−10	−15	−0	13
15	15	21	24	25	5	−4	28	29	9	−8	30	12	−11	14	−13	−1
ine																

Table 2. Multiplication rules of Kaluza numbers for $e_0, e_1, \ldots, e_{15}$ and $e_{16}, e_{17}, \ldots, e_{31}$ (elements e_i denoted by their subscripts, i.e., $i = e_i$).

ine×	16	17	18	19	20	21	22	23	24	25	26	27	28	29	30	31
*ine*0	16	17	18	19	20	21	22	23	24	25	26	27	28	29	30	31
1	10	11	12	13	14	15	26	27	28	29	22	23	24	25	31	30
2	−7	−8	−9	−26	−27	−28	−13	14	15	30	−19	−20	−21	−31	25	−29
3	−6	26	27	8	9	−29	11	12	−30	−15	−17	−18	31	21	24	−28
4	−26	−6	28	−7	29	9	−10	30	12	14	16	−31	−18	−20	−23	27
5	−27	−28	−6	−29	−7	−8	−30	−10	−11	−13	31	16	17	19	22	−26
6	−3	−4	−5	−22	−23	−24	19	20	21	31	−13	−14	−15	−30	29	−25
7	−2	22	23	4	5	−25	17	18	−31	−21	−11	−12	30	15	28	−24
8	−22	−2	24	−3	25	5	−16	31	18	20	10	−30	−12	−14	−27	23
9	−23	−24	−2	−25	−3	−4	−31	−16	−17	−19	30	10	11	13	26	−22
10	1	−19	−20	−17	−18	31	4	5	−25	−24	8	9	−29	−28	15	21
11	19	1	−21	16	−31	−18	−3	25	5	23	−7	29	9	27	−14	−20
12	20	21	1	31	16	17	−25	−3	−4	−22	−29	−7	−8	−26	13	19
13	17	−16	31	−1	21	−20	−2	24	−23	−5	−6	28	−27	−9	−12	−18
14	18	−31	−16	−21	−1	19	−24	−2	22	4	−28	−6	26	8	11	17
15	31	18	−17	20	−19	−1	23	−22	−2	−3	27	−26	−6	−7	−10	−16
ine																

Table 3. Multiplication rules of Kaluza numbers for $e_{16}, e_{17}, \ldots, e_{31}$ and $e_0, e_1, \ldots, e_{15}$ (elements e_i denoted by their subscripts, i.e., $i = e_i$).

ine×	0	1	2	3	4	5	6	7	8	9	10	11	12	13	14	15
*ine*16	16	10	−7	−6	26	27	−3	−2	22	23	1	−19	−20	−17	−18	31
17	17	11	−8	−26	−6	28	−4	−22	−2	24	19	1	−21	16	−31	−18
18	18	12	−9	−27	−28	−6	−5	−23	−24	−2	20	21	1	31	16	17
19	19	13	26	8	−7	29	22	4	−3	25	17	−16	31	−1	21	−20
20	20	14	27	9	−29	−7	23	5	−25	−3	18	−31	−16	−21	−1	19
21	21	15	28	29	9	−8	24	25	5	−4	31	18	−17	20	−19	−1
22	22	−26	13	11	−10	30	−19	−17	16	−31	4	−3	25	−2	24	−23
23	23	−27	14	12	−30	−10	−20	−18	31	16	5	−25	−3	−24	−2	22
24	24	−28	15	30	12	−11	−21	−31	−18	17	25	5	−4	23	−22	−2
25	25	−29	−30	−15	14	−13	31	21	−20	19	24	−23	22	−5	4	−3
26	26	−22	19	17	−16	31	−13	−11	10	−30	8	−7	29	−6	28	−27
27	27	−23	20	18	−31	−16	−14	−12	30	10	9	−29	−7	−28	−6	26
28	28	−24	21	31	18	−17	−15	−30	−12	11	29	9	−8	27	−26	−6
29	29	−25	−31	−21	20	−19	30	15	−14	13	28	−27	26	−9	8	−7
30	30	31	−25	−24	23	−22	−29	−28	27	−26	15	−14	13	−12	11	−10
31	31	30	−29	−28	27	−26	−25	−24	23	−22	21	−20	19	−18	17	−16
ine																

Table 4. Multiplication rules of Kaluza numbers for $e_{16}, e_{17}, \ldots, e_{31}$ and $e_{16}, e_{17}, \ldots, e_{31}$ (elements e_i denoted by their subscripts, i.e., $i = e_i$).

ine×	16	17	18	19	20	21	22	23	24	25	26	27	28	29	30	31
*ine*16	0	−13	−14	−11	−12	30	8	9	−29	−28	4	5	−25	−24	21	15
17	13	0	15	10	−30	−12	−7	29	9	27	−3	25	5	23	−20	−14
18	14	15	0	30	10	11	−29	−7	−8	−26	−25	−3	−4	−22	19	13
19	11	−10	30	−0	15	−14	−6	28	−27	−9	−2	24	−23	−5	−18	−12
20	12	−30	−10	−15	−0	13	−28	−6	26	8	−24	−2	22	4	17	11
21	30	12	−11	14	−13	−0	27	−26	−6	−7	23	−22	−2	−3	−16	−10
22	−8	7	−29	6	−28	27	−0	15	−14	−12	1	−21	20	18	−5	9
23	−9	29	7	28	6	−26	−15	−0	13	11	21	1	−19	−17	4	−8
24	−29	−9	8	−27	26	6	14	−13	−0	−10	−20	19	1	16	−3	7
25	−28	27	−26	9	−8	7	12	−11	10	0	−18	17	−16	−1	−2	6
26	−4	3	−25	2	−24	23	−1	21	−20	−18	0	−15	14	12	−9	5
27	−5	25	3	24	2	−22	−21	−1	19	17	15	0	−13	−11	8	−4
28	−25	−5	4	−23	22	2	20	−19	−1	−16	−14	13	0	10	−7	3
29	−24	23	−22	5	−4	3	18	−17	16	1	−12	11	−10	−0	−6	2
30	21	−20	19	−18	17	−16	5	−4	3	2	9	−8	7	6	−0	−1
31	15	−14	13	−12	11	−10	9	−8	7	6	5	−4	3	2	−1	−1
ine																

Suppose we want to compute the product of two Kaluza numbers:

$$d = d^{(1)}d^{(2)} = d_0 + \sum_{n=1}^{31} d_n e_n,$$

where

$$d^{(1)} = a_0 + \sum_{n=1}^{31} a_n e_n \quad \text{and} \quad d^{(2)} = b_0 + \sum_{n=1}^{31} b_n e_n.$$

The operation of the multiplication of Kaluza numbers can be represented more compactly in the form of a matrix-vector product:

$$\mathbf{Y}_{32\times1} = \mathbf{B}_{32}\mathbf{X}_{32\times1}, \tag{1}$$

where $\mathbf{Y}_{32\times1} = [d_0, d_1, \ldots, d_{31}]^{\mathrm{T}}$, $\mathbf{X}_{32\times1} = [a_0, a_1, \ldots, a_{31}]^{\mathrm{T}}$,

$$\mathbf{B}_{32} = \begin{bmatrix} \mathbf{B}_{16}^{(0,0)} & \mathbf{B}_{16}^{(1,0)} \\ \mathbf{B}_{16}^{(0,1)} & \mathbf{B}_{16}^{(1,1)} \end{bmatrix},$$

$$\mathbf{B}_{16}^{(0,0)} = \begin{bmatrix}
b_0 & b_1 & b_2 & -b_3 & -b_4 & -b_5 & -b_6 & b_7 & b_8 & b_9 & b_{10} & b_{11} & b_{12} & -b_{13} & -b_{14} & -b_{15} \\
b_1 & b_0 & -b_6 & b_7 & b_8 & b_9 & b_2 & -b_3 & -b_4 & -b_5 & b_{16} & b_{17} & b_{18} & -b_{19} & -b_{20} & -b_{21} \\
b_2 & b_6 & b_0 & b_{10} & b_{11} & b_{12} & -b_1 & -b_{16} & -b_{17} & -b_{18} & -b_3 & -b_4 & -b_5 & -b_{22} & -b_{23} & -b_{24} \\
b_3 & b_7 & b_{10} & b_0 & b_{13} & b_{14} & -b_{16} & -b_1 & -b_{19} & -b_{20} & -b_2 & -b_{22} & -b_{23} & -b_4 & -b_5 & -b_{25} \\
b_4 & b_8 & b_{11} & -b_{13} & b_0 & b_{15} & -b_{17} & b_{19} & -b_1 & -b_{21} & b_{22} & -b_2 & -b_{24} & b_3 & b_{25} & -b_5 \\
b_5 & b_9 & b_{12} & -b_{14} & -b_{15} & b_0 & -b_{18} & b_{20} & b_{21} & -b_1 & b_{23} & b_{24} & -b_2 & -b_{25} & b_3 & b_4 \\
b_6 & b_2 & -b_1 & -b_{16} & -b_{17} & -b_{18} & b_0 & b_{10} & b_{11} & b_{12} & -b_7 & -b_8 & -b_9 & -b_{26} & -b_{27} & -b_{28} \\
b_7 & b_3 & -b_{16} & -b_1 & -b_{19} & -b_{20} & b_{10} & b_0 & b_{13} & b_{14} & -b_6 & -b_{26} & -b_{27} & -b_8 & -b_9 & -b_{29} \\
b_8 & b_4 & -b_{17} & b_{19} & -b_1 & -b_{21} & b_{11} & -b_{13} & b_0 & b_{15} & b_{26} & -b_6 & -b_{28} & b_7 & b_{29} & -b_9 \\
b_9 & b_5 & -b_{18} & b_{20} & b_{21} & -b_1 & b_{12} & -b_{14} & -b_{15} & b_0 & b_{27} & b_{28} & -b_6 & -b_{29} & b_7 & b_8 \\
b_{10} & b_{16} & b_3 & -b_2 & -b_{22} & -b_{23} & -b_7 & b_6 & b_{26} & b_{27} & b_0 & b_{13} & b_{14} & -b_{11} & -b_{12} & -b_{30} \\
b_{11} & b_{17} & b_4 & b_{22} & -b_2 & -b_{24} & -b_8 & -b_{26} & b_6 & b_{28} & -b_{13} & b_0 & b_{15} & b_{10} & b_{30} & -b_{12} \\
b_{12} & b_{18} & b_5 & b_{23} & b_{24} & -b_2 & -b_9 & -b_{27} & -b_{28} & b_6 & -b_{14} & -b_{15} & b_0 & -b_{30} & b_{10} & b_{11} \\
b_{13} & b_{19} & b_{22} & b_4 & -b_3 & -b_{25} & -b_{26} & -b_8 & b_7 & b_{29} & -b_{11} & b_{10} & b_{30} & b_0 & b_{15} & -b_{14} \\
b_{14} & b_{20} & b_{23} & b_5 & b_{25} & -b_3 & -b_{27} & -b_9 & -b_{29} & b_7 & -b_{12} & -b_{30} & b_{10} & -b_{15} & b_0 & b_{13} \\
b_{15} & b_{21} & b_{24} & -b_{25} & b_5 & -b_4 & -b_{28} & b_{29} & -b_9 & b_8 & b_{30} & -b_{12} & b_{11} & b_{14} & -b_{13} & b_0
\end{bmatrix},$$

$$\mathbf{B}_{16}^{(1,0)} = \begin{bmatrix}
b_{16} & b_{10} & -b_7 & b_6 & b_{26} & b_{27} & b_3 & -b_2 & -b_{22} & -b_{23} & b_1 & b_{19} & b_{20} & -b_{17} & -b_{18} & -b_{31} \\
b_{17} & b_{11} & -b_8 & -b_{26} & b_6 & b_{28} & b_4 & b_{22} & -b_2 & -b_{24} & -b_{19} & b_1 & b_{21} & b_{16} & b_{31} & -b_{18} \\
b_{18} & b_{12} & -b_9 & -b_{27} & -b_{28} & b_6 & b_5 & b_{23} & b_{24} & -b_2 & -b_{20} & -b_{21} & b_1 & -b_{31} & b_{16} & b_{17} \\
b_{19} & b_{13} & -b_{26} & -b_8 & b_7 & b_{29} & b_{22} & b_4 & -b_3 & -b_{25} & -b_{17} & b_{16} & b_{31} & b_1 & b_{21} & -b_{20} \\
b_{20} & b_{14} & -b_{27} & -b_9 & -b_{29} & b_7 & b_{23} & b_5 & b_{25} & -b_3 & -b_{18} & -b_{31} & b_{16} & -b_{21} & b_1 & b_{19} \\
b_{21} & b_{15} & -b_{28} & b_{29} & -b_9 & b_8 & b_{24} & -b_{25} & b_5 & -b_4 & b_{31} & -b_{18} & b_{17} & b_{20} & -b_{19} & b_1 \\
b_{22} & b_{26} & b_{13} & -b_{11} & b_{10} & b_{30} & -b_{19} & b_{17} & -b_{16} & -b_{31} & b_4 & -b_3 & -b_{25} & b_2 & b_{24} & -b_{23} \\
b_{23} & b_{27} & b_{14} & -b_{12} & -b_{30} & b_{10} & -b_{20} & b_{18} & b_{31} & -b_{16} & b_5 & b_{25} & -b_3 & -b_{24} & b_2 & b_{22} \\
b_{24} & b_{28} & b_{15} & b_{30} & -b_{12} & b_{11} & -b_{21} & -b_{31} & b_{18} & -b_{17} & -b_{25} & b_5 & -b_4 & b_{23} & -b_{22} & b_2 \\
b_{25} & b_{29} & b_{30} & b_{15} & -b_{14} & b_{13} & -b_{31} & -b_{21} & b_{20} & -b_{19} & -b_{24} & b_{23} & -b_{22} & b_5 & -b_4 & b_3 \\
b_{26} & b_{22} & -b_{19} & b_{17} & -b_{16} & -b_{31} & b_{13} & -b_{11} & b_{10} & b_{30} & b_8 & -b_7 & -b_{29} & b_6 & b_{28} & -b_{27} \\
b_{27} & b_{23} & -b_{20} & b_{18} & b_{31} & -b_{16} & b_{14} & -b_{12} & -b_{30} & b_{10} & b_9 & b_{29} & -b_7 & -b_{28} & b_6 & b_{26} \\
b_{28} & b_{24} & -b_{21} & -b_{31} & b_{18} & -b_{17} & b_{15} & b_{30} & -b_{12} & b_{11} & -b_{29} & b_9 & -b_8 & b_{27} & -b_{26} & b_6 \\
b_{29} & b_{25} & -b_{31} & -b_{21} & b_{20} & -b_{19} & b_{30} & b_{15} & -b_{14} & b_{13} & -b_{28} & b_{27} & -b_{26} & b_9 & -b_8 & b_7 \\
b_{30} & b_{31} & b_{25} & -b_{24} & b_{23} & -b_{22} & -b_{29} & b_{28} & -b_{27} & b_{26} & b_{15} & -b_{14} & b_{13} & b_{12} & -b_{11} & b_{10} \\
b_{31} & b_{30} & -b_{29} & b_{28} & -b_{27} & b_{26} & b_{25} & -b_{24} & b_{23} & -b_{22} & b_{21} & -b_{20} & b_{19} & b_{18} & -b_{17} & b_{16}
\end{bmatrix},$$

$$\mathbf{B}_{16}^{(0,1)} = \begin{bmatrix} b_{16} & b_{17} & b_{18} & -b_{19} & -b_{20} & -b_{21} & -b_{22} & -b_{23} & -b_{24} & b_{25} & b_{26} & b_{27} & b_{28} & -b_{29} & -b_{30} & -b_{31} \\ b_{10} & b_{11} & b_{12} & -b_{13} & -b_{14} & -b_{15} & b_{26} & b_{27} & b_{28} & -b_{29} & -b_{22} & -b_{23} & -b_{24} & b_{25} & -b_{31} & -b_{30} \\ -b_{7} & -b_{8} & -b_{9} & -b_{26} & -b_{27} & -b_{28} & -b_{13} & -b_{14} & -b_{15} & -b_{30} & b_{19} & b_{20} & b_{21} & b_{31} & b_{25} & b_{29} \\ -b_{6} & -b_{26} & -b_{27} & -b_{8} & -b_{9} & -b_{29} & -b_{11} & -b_{12} & -b_{30} & -b_{15} & b_{17} & b_{18} & b_{31} & b_{21} & b_{24} & b_{28} \\ b_{26} & -b_{6} & -b_{28} & b_{7} & b_{29} & -b_{9} & b_{10} & b_{30} & -b_{12} & b_{14} & -b_{16} & -b_{31} & b_{18} & -b_{20} & -b_{23} & -b_{27} \\ b_{27} & b_{28} & -b_{6} & -b_{29} & b_{7} & b_{8} & -b_{30} & b_{10} & b_{11} & -b_{13} & b_{31} & -b_{16} & -b_{17} & b_{19} & b_{22} & b_{26} \\ -b_{3} & -b_{4} & -b_{5} & -b_{22} & -b_{23} & -b_{24} & b_{19} & b_{20} & b_{21} & b_{31} & -b_{13} & -b_{14} & -b_{15} & -b_{30} & b_{29} & b_{25} \\ -b_{2} & -b_{22} & -b_{23} & -b_{4} & -b_{5} & -b_{25} & b_{17} & b_{18} & b_{31} & b_{21} & -b_{11} & -b_{12} & -b_{30} & -b_{15} & b_{28} & b_{24} \\ b_{22} & -b_{2} & -b_{24} & b_{3} & b_{25} & -b_{5} & -b_{16} & -b_{31} & b_{18} & -b_{20} & b_{10} & b_{30} & -b_{12} & b_{14} & -b_{27} & -b_{23} \\ b_{23} & b_{24} & -b_{2} & -b_{25} & b_{3} & b_{4} & b_{31} & -b_{16} & -b_{17} & b_{19} & -b_{30} & b_{10} & b_{11} & -b_{13} & b_{26} & b_{22} \\ b_{1} & b_{19} & b_{20} & -b_{17} & -b_{18} & -b_{31} & -b_{4} & -b_{5} & -b_{25} & b_{24} & b_{8} & b_{9} & b_{29} & -b_{28} & -b_{15} & -b_{21} \\ -b_{19} & b_{1} & b_{21} & b_{16} & b_{31} & -b_{18} & b_{3} & b_{25} & -b_{5} & -b_{23} & -b_{7} & -b_{29} & b_{9} & b_{27} & b_{14} & b_{20} \\ -b_{20} & -b_{21} & b_{1} & -b_{31} & b_{16} & b_{17} & -b_{25} & b_{3} & b_{4} & b_{22} & b_{29} & -b_{7} & -b_{8} & -b_{26} & -b_{13} & -b_{19} \\ -b_{17} & b_{16} & b_{31} & b_{1} & b_{21} & -b_{20} & b_{2} & b_{24} & -b_{23} & -b_{5} & -b_{6} & -b_{28} & b_{27} & b_{9} & b_{12} & b_{18} \\ -b_{18} & -b_{31} & b_{16} & -b_{21} & b_{1} & b_{19} & -b_{24} & b_{2} & b_{22} & b_{4} & b_{28} & -b_{6} & -b_{26} & -b_{8} & -b_{11} & -b_{17} \\ b_{31} & -b_{18} & b_{17} & b_{20} & -b_{19} & b_{1} & b_{23} & -b_{22} & b_{2} & -b_{3} & -b_{27} & b_{26} & -b_{6} & b_{7} & b_{10} & b_{16} \end{bmatrix},$$

$$\mathbf{B}_{16}^{(1,1)} = \begin{bmatrix} b_{0} & b_{13} & b_{14} & -b_{11} & -b_{12} & -b_{30} & b_{8} & b_{9} & b_{29} & -b_{28} & -b_{4} & -b_{5} & -b_{25} & b_{24} & -b_{21} & -b_{15} \\ -b_{13} & b_{0} & b_{15} & b_{10} & b_{30} & -b_{12} & -b_{7} & -b_{29} & b_{9} & b_{27} & b_{3} & b_{25} & -b_{5} & -b_{23} & b_{20} & b_{14} \\ -b_{14} & -b_{15} & b_{0} & -b_{30} & b_{10} & b_{11} & b_{29} & -b_{7} & -b_{8} & -b_{26} & -b_{25} & b_{3} & b_{4} & b_{22} & -b_{19} & -b_{13} \\ -b_{11} & b_{10} & b_{30} & b_{0} & b_{15} & -b_{14} & -b_{6} & -b_{28} & b_{27} & b_{9} & b_{2} & b_{24} & -b_{23} & -b_{5} & b_{18} & b_{12} \\ -b_{12} & -b_{30} & b_{10} & -b_{15} & b_{0} & b_{13} & b_{28} & -b_{6} & -b_{26} & -b_{8} & -b_{24} & b_{2} & b_{22} & b_{4} & -b_{17} & -b_{11} \\ b_{30} & -b_{12} & b_{11} & b_{14} & -b_{13} & b_{0} & -b_{27} & b_{26} & -b_{6} & b_{7} & b_{23} & -b_{22} & b_{2} & -b_{3} & b_{16} & b_{10} \\ b_{8} & -b_{7} & -b_{29} & b_{6} & b_{28} & -b_{27} & b_{0} & b_{15} & -b_{14} & b_{12} & -b_{1} & -b_{21} & b_{20} & -b_{18} & -b_{5} & -b_{9} \\ b_{9} & b_{29} & -b_{7} & -b_{28} & b_{6} & b_{26} & -b_{15} & b_{0} & b_{13} & -b_{11} & b_{21} & -b_{1} & -b_{19} & b_{17} & b_{4} & b_{8} \\ -b_{29} & b_{9} & -b_{8} & b_{27} & -b_{26} & b_{6} & b_{14} & -b_{13} & b_{0} & b_{10} & -b_{20} & b_{19} & -b_{1} & -b_{16} & -b_{3} & -b_{7} \\ -b_{28} & b_{27} & -b_{26} & b_{9} & -b_{8} & b_{7} & b_{12} & -b_{11} & b_{10} & b_{0} & -b_{18} & b_{17} & -b_{16} & -b_{1} & -b_{2} & -b_{6} \\ b_{4} & -b_{3} & -b_{25} & b_{2} & b_{24} & -b_{23} & -b_{1} & -b_{21} & b_{20} & -b_{18} & b_{0} & b_{15} & -b_{14} & b_{12} & -b_{9} & -b_{5} \\ b_{5} & b_{25} & -b_{3} & -b_{24} & b_{2} & b_{22} & b_{21} & -b_{1} & -b_{19} & b_{17} & -b_{15} & b_{0} & b_{13} & -b_{11} & b_{8} & b_{4} \\ -b_{25} & b_{5} & -b_{4} & b_{23} & -b_{22} & b_{2} & -b_{20} & b_{19} & -b_{1} & -b_{16} & b_{14} & -b_{13} & b_{0} & b_{10} & -b_{7} & -b_{3} \\ -b_{24} & b_{23} & -b_{22} & b_{5} & -b_{4} & b_{3} & -b_{18} & b_{17} & -b_{16} & -b_{1} & b_{12} & -b_{11} & b_{10} & b_{0} & -b_{6} & -b_{2} \\ b_{21} & -b_{20} & b_{19} & b_{18} & -b_{17} & b_{16} & b_{5} & -b_{4} & b_{3} & -b_{2} & -b_{9} & b_{8} & -b_{7} & b_{6} & b_{0} & b_{1} \\ b_{15} & -b_{14} & b_{13} & b_{12} & -b_{11} & b_{10} & -b_{9} & b_{8} & -b_{7} & b_{6} & b_{5} & -b_{4} & b_{3} & -b_{2} & b_{1} & b_{0} \end{bmatrix}.$$

The direct multiplication of the matrix-vector product in Equation (1) requires 1024 real multiplications and 992 additions. We shall present an algorithm that reduces computation complexity to 192 multiplications and 384 additions of real numbers.

3. Synthesis of a Rationalized Algorithm for Computing Kaluza Numbers Product

We first rearrange the rows and columns of the matrix respectively using the permutations: $\pi_r = (11, 17, 2, 6, 13, 19, 3, 7, 0, 1, 4, 8, 10, 16, 22, 26, 30, 31, 23, 27, 15, 21, 5, 9, 14, 20, 25, 29, 12, 18, 24, 28)$ and $\pi_c = (10, 16, 3, 7, 0, 1, 2, 6, 13, 19, 22, 26, 11, 17, 4, 8, 12, 18, 5, 9, 14, 20, 23, 27, 15, 21, 24, 28, 30, 31, 25, 29)$. Next, we change the sign of the selected rows $\{8, 9, 12, 13, 14, 15, 26, 27\}$ and columns $\{2, 3, 6, 7, 8, 9, 12, 13\}$ by multiplying them by -1. We can easily see that this transformation leads in the future to minimizing the computational complexity of the final algorithm. Then we can write:

$$\mathbf{Y}_{32\times 1} = \mathbf{M}_{32}^{(r)} \breve{\mathbf{B}}_{32} \mathbf{M}_{32}^{(c)} \mathbf{X}_{32\times 1}, \tag{2}$$

where the monomial matrices $\mathbf{M}_{32}^{(r)}$, $\mathbf{M}_{32}^{(c)}$ are products of an appriopriate alternating sign changing matrices $\mathbf{S}_{32}^{(r)}$, $\mathbf{S}_{32}^{(c)}$ and a permutation matrix $\mathbf{P}_{32}^{(r)}$, $\mathbf{P}_{32}^{(c)}$:

$$\mathbf{M}_{32}^{(r)} = \mathbf{S}_{32}^{(r)} \mathbf{P}_{32}^{(r)},$$

$$\mathbf{M}_{32}^{(c)} = \mathbf{P}_{32}^{(c)} \mathbf{S}_{32}^{(c)},$$

where:

$$\mathbf{P}_{32}^{(c)} = \begin{bmatrix} \mathbf{P}_{16}^{(c,(0,0))} & \mathbf{P}_{16}^{(c,(0,1))} \\ \mathbf{P}_{16}^{(c,(1,0))} & \mathbf{P}_{16}^{(c,(1,1))} \end{bmatrix},$$

$$\mathbf{P}_{16}^{(c(0,0))} = \begin{bmatrix}
0&0&0&0&0&0&0&0&0&0&0&1&0&0&0&0\\
0&0&0&0&0&0&0&0&0&0&0&0&0&0&0&0\\
0&0&1&0&0&0&0&0&0&0&0&0&0&0&0&0\\
0&0&0&0&0&0&1&0&0&0&0&0&0&0&0&0\\
0&0&0&0&0&0&0&0&0&0&0&0&0&1&0&0\\
0&0&0&0&0&0&0&0&0&0&0&0&0&0&0&0\\
0&0&0&1&0&0&0&0&0&0&0&0&0&0&0&0\\
0&0&0&0&0&0&0&1&0&0&0&0&0&0&0&0\\
1&0&0&0&0&0&0&0&0&0&0&0&0&0&0&0\\
0&1&0&0&0&0&0&0&0&0&0&0&0&0&0&0\\
0&0&0&0&1&0&0&0&0&0&0&0&0&0&0&0\\
0&0&0&0&0&0&0&0&1&0&0&0&0&0&0&0\\
0&0&0&0&0&0&0&0&0&0&1&0&0&0&0&0\\
0&0&0&0&0&0&0&0&0&0&0&0&0&0&0&0\\
0&0&0&0&0&0&0&0&0&0&0&0&0&0&0&0\\
0&0&0&0&0&0&0&0&0&0&0&0&0&0&0&0
\end{bmatrix},$$

$$\mathbf{P}_{16}^{(c(0,1))} = \begin{bmatrix}
0&0&0&0&0&0&0&0&0&0&0&0&0&0&0&0\\
0&1&0&0&0&0&0&0&0&0&0&0&0&0&0&0\\
0&0&0&0&0&0&0&0&0&0&0&0&0&0&0&0\\
0&0&0&0&0&0&0&0&0&0&0&0&0&0&0&0\\
0&0&0&0&0&0&0&0&0&0&0&0&0&0&0&0\\
0&0&0&1&0&0&0&0&0&0&0&0&0&0&0&0\\
0&0&0&0&0&0&0&0&0&0&0&0&0&0&0&0\\
0&0&0&0&0&0&0&0&0&0&0&0&0&0&0&0\\
0&0&0&0&0&0&0&0&0&0&0&0&0&0&0&0\\
0&0&0&0&0&0&0&0&0&0&0&0&0&0&0&0\\
0&0&0&0&0&0&0&0&0&0&0&0&0&0&0&0\\
0&0&0&0&0&0&0&0&0&0&0&0&0&0&0&0\\
0&0&0&0&0&0&0&0&0&0&0&0&0&0&0&0\\
1&0&0&0&0&0&0&0&0&0&0&0&0&0&0&0\\
0&0&0&0&0&0&1&0&0&0&0&0&0&0&0&0\\
0&0&0&0&0&0&0&0&0&0&1&0&0&0&0&0
\end{bmatrix},$$

$$\mathbf{P}_{16}^{(c(1,0))} = \begin{bmatrix}
0&0&0&0&0&0&0&0&0&0&0&0&0&0&0&0\\
0&0&0&0&0&0&0&0&0&0&0&0&0&0&0&0\\
0&0&0&0&0&0&0&0&0&0&0&0&0&0&0&0\\
0&0&0&0&0&0&0&0&0&0&0&0&0&0&0&0\\
0&0&0&0&0&0&0&0&0&0&0&0&0&0&0&1\\
0&0&0&0&0&0&0&0&0&0&0&0&0&0&0&0\\
0&0&0&0&0&1&0&0&0&0&0&0&0&0&0&0\\
0&0&0&0&0&0&0&0&0&1&0&0&0&0&0&0\\
0&0&0&0&0&0&0&0&0&0&0&0&0&0&1&0\\
0&0&0&0&0&0&0&0&0&0&0&0&0&0&0&0\\
0&0&0&0&0&0&0&0&0&0&0&0&0&0&0&0\\
0&0&0&0&0&0&0&0&0&0&0&0&0&0&0&0\\
0&0&0&0&0&0&0&0&0&0&0&0&1&0&0&0\\
0&0&0&0&0&0&0&0&0&0&0&0&0&0&0&0\\
0&0&0&0&0&0&0&0&0&0&0&0&0&0&0&0\\
0&0&0&0&0&0&0&0&0&0&0&0&0&0&0&0
\end{bmatrix},$$

$$\mathbf{P}_{16}^{(c(1,1))} = \begin{bmatrix}
0 & 0 & 0 & 0 & 0 & 0 & 0 & 0 & 0 & 0 & 0 & 0 & 0 & 0 & 1 & 0 \\
0 & 0 & 0 & 0 & 0 & 0 & 0 & 0 & 0 & 0 & 0 & 0 & 0 & 0 & 0 & 1 \\
0 & 0 & 0 & 0 & 0 & 0 & 0 & 1 & 0 & 0 & 0 & 0 & 0 & 0 & 0 & 0 \\
0 & 0 & 0 & 0 & 0 & 0 & 0 & 0 & 0 & 0 & 0 & 1 & 0 & 0 & 0 & 0 \\
0 & 0 & 0 & 0 & 0 & 0 & 0 & 0 & 0 & 0 & 0 & 0 & 0 & 0 & 0 & 0 \\
0 & 0 & 0 & 0 & 0 & 1 & 0 & 0 & 0 & 0 & 0 & 0 & 0 & 0 & 0 & 0 \\
0 & 0 & 0 & 0 & 0 & 0 & 0 & 0 & 0 & 0 & 0 & 0 & 0 & 0 & 0 & 0 \\
0 & 0 & 0 & 0 & 0 & 0 & 0 & 0 & 0 & 0 & 0 & 0 & 0 & 0 & 0 & 0 \\
0 & 0 & 0 & 0 & 0 & 0 & 0 & 0 & 0 & 0 & 0 & 0 & 0 & 0 & 0 & 0 \\
0 & 0 & 0 & 0 & 1 & 0 & 0 & 0 & 0 & 0 & 0 & 0 & 0 & 0 & 0 & 0 \\
0 & 0 & 0 & 0 & 0 & 0 & 0 & 0 & 0 & 1 & 0 & 0 & 0 & 0 & 0 & 0 \\
0 & 0 & 0 & 0 & 0 & 0 & 0 & 0 & 0 & 0 & 0 & 0 & 0 & 1 & 0 & 0 \\
0 & 0 & 0 & 0 & 0 & 0 & 0 & 0 & 0 & 0 & 0 & 0 & 0 & 0 & 0 & 0 \\
0 & 0 & 1 & 0 & 0 & 0 & 0 & 0 & 0 & 0 & 0 & 0 & 0 & 0 & 0 & 0 \\
0 & 0 & 0 & 0 & 0 & 0 & 0 & 0 & 1 & 0 & 0 & 0 & 0 & 0 & 0 & 0 \\
0 & 0 & 0 & 0 & 0 & 0 & 0 & 0 & 0 & 0 & 0 & 0 & 1 & 0 & 0 & 0
\end{bmatrix},$$

$$\mathbf{S}_{\mathbf{32}}^{(\mathbf{c})} = \mathrm{diag}\big([-1 \;\; -1 \;\; -1 \;\; -1 \;\; 1 \;\; 1 \;\; -1 \;\; -1 \;\; 1 \;\; 1 \;\; -1 \;\; 1 \;\; 1 \;\; 1 \;\; 1 \;\; 1 \;\; -1 \;\; 1 \;\; 1 \;\; 1 \;\; 1 \;\; 1 \;\; 1 \;\; 1 \;\; 1 \;\; 1 \;\; 1 \;\; 1 \;\; 1 \;\; 1 \;\; 1 \;\; 1]\big),$$

$$\mathbf{P}_{32}^{(r)} = \begin{bmatrix} \mathbf{P}_{16}^{(r,(0,0))} & \mathbf{P}_{16}^{(r,(0,1))} \\ \mathbf{P}_{16}^{(r,(1,0))} & \mathbf{P}_{16}^{(r,(1,1))} \end{bmatrix},$$

$$\mathbf{P}_{16}^{(r,(0,0))} = \begin{bmatrix}
0 & 0 & 0 & 0 & 1 & 0 & 0 & 0 & 0 & 0 & 0 & 0 & 0 & 0 & 0 & 0 \\
0 & 0 & 0 & 0 & 0 & 1 & 0 & 0 & 0 & 0 & 0 & 0 & 0 & 0 & 0 & 0 \\
0 & 0 & 0 & 0 & 0 & 0 & 1 & 0 & 0 & 0 & 0 & 0 & 0 & 0 & 0 & 0 \\
0 & 0 & 1 & 0 & 0 & 0 & 0 & 0 & 0 & 0 & 0 & 0 & 0 & 0 & 0 & 0 \\
0 & 0 & 0 & 0 & 0 & 0 & 0 & 0 & 0 & 0 & 0 & 0 & 0 & 0 & 1 & 0 \\
0 & 0 & 0 & 0 & 0 & 0 & 0 & 0 & 0 & 0 & 0 & 0 & 0 & 0 & 0 & 0 \\
0 & 0 & 0 & 0 & 0 & 0 & 0 & 1 & 0 & 0 & 0 & 0 & 0 & 0 & 0 & 0 \\
0 & 0 & 0 & 1 & 0 & 0 & 0 & 0 & 0 & 0 & 0 & 0 & 0 & 0 & 0 & 0 \\
0 & 0 & 0 & 0 & 0 & 0 & 0 & 0 & 0 & 0 & 0 & 0 & 0 & 0 & 0 & 1 \\
0 & 0 & 0 & 0 & 0 & 0 & 0 & 0 & 0 & 0 & 0 & 0 & 0 & 0 & 0 & 0 \\
1 & 0 & 0 & 0 & 0 & 0 & 0 & 0 & 0 & 0 & 0 & 0 & 0 & 0 & 0 & 0 \\
0 & 0 & 0 & 0 & 0 & 0 & 0 & 0 & 0 & 0 & 0 & 0 & 1 & 0 & 0 & 0 \\
0 & 0 & 0 & 0 & 0 & 0 & 0 & 0 & 0 & 0 & 0 & 0 & 0 & 0 & 0 & 0 \\
0 & 0 & 0 & 0 & 0 & 0 & 0 & 0 & 1 & 0 & 0 & 0 & 0 & 0 & 0 & 0 \\
0 & 0 & 0 & 0 & 0 & 0 & 0 & 0 & 0 & 0 & 0 & 0 & 0 & 0 & 0 & 0 \\
0 & 0 & 0 & 0 & 0 & 0 & 0 & 0 & 0 & 0 & 0 & 0 & 0 & 0 & 0 & 0
\end{bmatrix},$$

$$\mathbf{P}_{16}^{(r,(0,1))} = \begin{bmatrix}
0 & 0 & 0 & 0 & 0 & 0 & 0 & 0 & 0 & 0 & 0 & 0 & 0 & 0 & 0 & 0 \\
0 & 0 & 0 & 0 & 0 & 0 & 0 & 0 & 0 & 0 & 0 & 0 & 0 & 0 & 0 & 0 \\
0 & 0 & 0 & 0 & 0 & 0 & 0 & 0 & 0 & 0 & 0 & 0 & 0 & 0 & 0 & 0 \\
0 & 0 & 0 & 0 & 0 & 0 & 0 & 0 & 0 & 0 & 0 & 0 & 0 & 0 & 0 & 0 \\
0 & 0 & 0 & 0 & 0 & 0 & 0 & 0 & 0 & 0 & 0 & 0 & 0 & 0 & 0 & 0 \\
0 & 0 & 1 & 0 & 0 & 0 & 0 & 0 & 0 & 0 & 0 & 0 & 0 & 0 & 0 & 0 \\
0 & 0 & 0 & 0 & 0 & 0 & 0 & 0 & 0 & 0 & 0 & 0 & 0 & 0 & 0 & 0 \\
0 & 0 & 0 & 0 & 0 & 0 & 0 & 0 & 0 & 0 & 0 & 0 & 0 & 0 & 0 & 0 \\
0 & 0 & 0 & 0 & 0 & 0 & 0 & 0 & 0 & 0 & 0 & 0 & 0 & 0 & 0 & 0 \\
0 & 0 & 0 & 1 & 0 & 0 & 0 & 0 & 0 & 0 & 0 & 0 & 0 & 0 & 0 & 0 \\
0 & 0 & 0 & 0 & 0 & 0 & 0 & 0 & 0 & 0 & 0 & 0 & 0 & 0 & 0 & 0 \\
0 & 0 & 0 & 0 & 0 & 0 & 0 & 0 & 0 & 0 & 0 & 0 & 0 & 0 & 0 & 0 \\
1 & 0 & 0 & 0 & 0 & 0 & 0 & 0 & 0 & 0 & 0 & 0 & 0 & 0 & 0 & 0 \\
0 & 0 & 0 & 0 & 0 & 0 & 0 & 0 & 0 & 0 & 0 & 0 & 0 & 0 & 0 & 0 \\
0 & 0 & 0 & 0 & 1 & 0 & 0 & 0 & 0 & 0 & 0 & 0 & 0 & 0 & 0 & 0 \\
0 & 0 & 0 & 0 & 0 & 0 & 0 & 0 & 1 & 0 & 0 & 0 & 0 & 0 & 0 & 0
\end{bmatrix},$$

$$\mathbf{P}_{16}^{(r,(1,0))} = \begin{bmatrix} 0&1&0&0&0&0&0&0&0&0&0&0&0&0&0&0 \\ 0&0&0&0&0&0&0&0&0&0&0&0&0&1&0&0 \\ 0&0&0&0&0&0&0&0&0&0&0&0&0&0&0&0 \\ 0&0&0&0&0&0&0&0&0&1&0&0&0&0&0&0 \\ 0&0&0&0&0&0&0&0&0&0&0&0&0&0&0&0 \\ 0&0&0&0&0&0&0&0&0&0&0&0&0&0&0&0 \\ 0&0&0&0&0&0&0&0&0&0&1&0&0&0&0&0 \\ 0&0&0&0&0&0&0&0&0&0&0&0&0&0&0&0 \\ 0&0&0&0&0&0&0&0&0&0&0&0&0&0&0&0 \\ 0&0&0&0&0&0&0&0&0&0&0&0&0&0&0&0 \\ 0&0&0&0&0&0&0&0&0&0&0&1&0&0&0&0 \\ 0&0&0&0&0&0&0&0&0&0&0&0&0&0&0&0 \\ 0&0&0&0&0&0&0&0&0&0&0&0&0&0&0&0 \\ 0&0&0&0&0&0&0&0&0&0&0&0&0&0&0&0 \\ 0&0&0&0&0&0&0&0&0&0&0&0&0&0&0&0 \\ 0&0&0&0&0&0&0&0&0&0&0&0&0&0&0&0 \end{bmatrix},$$

$$\mathbf{P}_{16}^{(r,(1,1))} = \begin{bmatrix} 0&0&0&0&0&0&0&0&0&0&0&0&0&0&0&0 \\ 0&0&0&0&0&0&0&0&0&0&0&0&0&0&0&0 \\ 0&1&0&0&0&0&0&0&0&0&0&0&0&0&0&0 \\ 0&0&0&0&0&0&0&0&0&0&0&0&0&0&0&0 \\ 0&0&0&0&0&1&0&0&0&0&0&0&0&0&0&0 \\ 0&0&0&0&0&0&0&0&0&1&0&0&0&0&0&0 \\ 0&0&0&0&0&0&0&0&0&0&0&0&0&0&0&0 \\ 0&0&0&0&0&0&1&0&0&0&0&0&0&0&0&0 \\ 0&0&0&0&0&0&0&0&0&0&1&0&0&0&0&0 \\ 0&0&0&0&0&0&0&0&0&0&0&0&0&0&1&0 \\ 0&0&0&0&0&0&0&0&0&0&0&0&0&0&0&0 \\ 0&0&0&0&0&0&0&1&0&0&0&0&0&0&0&0 \\ 0&0&0&0&0&0&0&0&0&0&0&1&0&0&0&0 \\ 0&0&0&0&0&0&0&0&0&0&0&0&0&0&0&1 \\ 0&0&0&0&0&0&0&0&0&0&0&0&1&0&0&0 \\ 0&0&0&0&0&0&0&0&0&0&0&0&0&1&0&0 \end{bmatrix},$$

$$\mathbf{S}_{32}^{(\mathrm{r})} = \mathrm{diag}(1,\ 1,\ 1,\ 1,\ -1,\ 1,\ 1,\ 1,\ -1,\ 1,\ 1,\ -1,\ 1,\ -1,\ 1,\ 1,\ 1, \\ -1,\ 1,\ -1,\ 1,\ 1,\ 1,\ 1,\ -1,\ 1,\ 1,\ 1,\ -1,\ 1,\ 1,\ 1).$$

The matrix $\breve{\mathbf{B}}_{32}$ is calculated from:

$$\breve{\mathbf{B}}_{32} = \left(\mathbf{M}_{32}^{(r)}\right)^{-1} \mathbf{B}_{32} \left(\mathbf{M}_{32}^{(c)}\right)^{-1},$$

If we interpret the $\breve{\mathbf{B}}_{32}$ matrix as a block matrix, it is easy to see that it has a bisymmetric structure:

$$\breve{\mathbf{B}}_{32} = \begin{bmatrix} \breve{\mathbf{B}}_{16}^{(0)} & \breve{\mathbf{B}}_{16}^{(1)} \\ \breve{\mathbf{B}}_{16}^{(1)} & -\breve{\mathbf{B}}_{16}^{(0)} \end{bmatrix},$$

where

$$\breve{\mathbf{B}}_{16}^{(0)} = \begin{bmatrix}
b_{13} & b_{19} & -b_3 & b_7 & -b_{11} & -b_{17} & b_2 & -b_6 & -b_{10} & -b_{16} & -b_{22} & b_{26} & -b_0 & -b_1 & -b_4 & b_8 \\
b_{19} & b_{13} & b_7 & -b_3 & -b_{17} & -b_{11} & -b_6 & b_2 & -b_{16} & -b_{10} & b_{26} & -b_{22} & -b_1 & -b_0 & b_8 & -b_4 \\
-b_{22} & -b_{26} & -b_{10} & b_{16} & -b_4 & -b_8 & -b_0 & b_1 & -b_3 & -b_7 & b_{13} & -b_{19} & b_2 & b_6 & -b_{11} & b_{17} \\
-b_{26} & -b_{22} & b_{16} & -b_{10} & -b_8 & -b_4 & b_1 & -b_0 & -b_7 & -b_3 & -b_{19} & b_{13} & b_6 & b_2 & b_{17} & -b_{11} \\
b_{11} & b_{17} & -b_2 & b_6 & -b_{13} & -b_{19} & b_3 & -b_7 & -b_0 & -b_1 & -b_4 & b_8 & -b_{10} & -b_{16} & -b_{22} & b_{26} \\
b_{17} & b_{11} & b_6 & -b_2 & -b_{19} & -b_{13} & -b_7 & b_3 & -b_1 & -b_0 & b_8 & -b_4 & -b_{16} & -b_{10} & b_{26} & -b_{22} \\
-b_4 & -b_8 & -b_0 & b_1 & -b_{22} & -b_{26} & -b_{10} & b_{16} & -b_2 & -b_6 & b_{11} & -b_{17} & b_3 & b_7 & -b_{13} & b_{19} \\
-b_8 & -b_4 & b_1 & -b_0 & -b_{26} & -b_{22} & b_{16} & -b_{10} & -b_6 & -b_2 & -b_{17} & b_{11} & b_7 & b_3 & b_{19} & -b_{13} \\
-b_{10} & -b_{16} & b_{22} & -b_{26} & -b_0 & -b_1 & b_4 & -b_8 & b_{13} & b_{19} & b_3 & -b_7 & -b_{11} & -b_{17} & -b_2 & b_6 \\
-b_{16} & -b_{10} & -b_{26} & b_{22} & -b_1 & -b_0 & -b_8 & b_4 & b_{19} & b_{13} & -b_7 & b_3 & -b_{17} & -b_{11} & b_6 & -b_2 \\
-b_3 & -b_7 & -b_{13} & b_{19} & b_2 & b_6 & b_{11} & -b_{17} & -b_{22} & -b_{26} & b_{10} & -b_{16} & -b_4 & -b_8 & b_0 & -b_1 \\
-b_7 & -b_3 & b_{19} & -b_{13} & b_6 & b_2 & -b_{17} & b_{11} & -b_{26} & -b_{22} & -b_{16} & b_{10} & -b_8 & -b_4 & -b_1 & b_0 \\
-b_0 & -b_1 & b_4 & -b_8 & -b_{10} & -b_{16} & b_{22} & -b_{26} & b_{11} & b_{17} & b_2 & -b_6 & -b_{13} & -b_{19} & -b_3 & b_7 \\
-b_1 & -b_0 & -b_8 & b_4 & -b_{16} & -b_{10} & -b_{26} & b_{22} & b_{17} & b_{11} & -b_6 & b_2 & -b_{19} & -b_{13} & b_7 & -b_3 \\
b_2 & b_6 & b_{11} & -b_{17} & -b_3 & -b_7 & -b_{13} & b_{19} & b_4 & b_8 & -b_0 & b_1 & b_{22} & b_{26} & -b_{10} & b_{16} \\
b_6 & b_2 & -b_{17} & b_{11} & -b_7 & -b_3 & b_{19} & -b_{13} & b_8 & b_4 & b_1 & -b_0 & b_{26} & b_{22} & b_{16} & -b_{10}
\end{bmatrix},$$

$$\breve{\mathbf{B}}_{16}^{(1)} = \begin{bmatrix}
-b_{15} & -b_{21} & -b_5 & b_9 & -b_{30} & -b_{31} & -b_{23} & b_{27} & -b_{12} & -b_{18} & b_{24} & -b_{28} & b_{14} & b_{20} & -b_{25} & b_{29} \\
-b_{21} & -b_{15} & b_9 & -b_5 & -b_{31} & -b_{30} & b_{27} & -b_{23} & -b_{18} & -b_{12} & -b_{28} & b_{24} & b_{20} & b_{14} & b_{29} & -b_{25} \\
b_{24} & b_{28} & -b_{12} & b_{18} & -b_{25} & -b_{29} & b_{14} & -b_{20} & -b_5 & -b_9 & -b_{15} & b_{21} & -b_{23} & -b_{27} & -b_{30} & b_{31} \\
b_{28} & b_{24} & b_{18} & -b_{12} & -b_{29} & -b_{25} & -b_{20} & b_{14} & -b_9 & -b_5 & b_{21} & -b_{15} & -b_{27} & -b_{23} & b_{31} & -b_{30} \\
-b_{30} & -b_{31} & -b_{23} & b_{27} & -b_{15} & -b_{21} & -b_5 & b_9 & -b_{14} & -b_{20} & b_{25} & -b_{29} & b_{12} & b_{18} & -b_{24} & b_{28} \\
-b_{31} & -b_{30} & b_{27} & -b_{23} & -b_{21} & -b_{15} & b_9 & -b_5 & -b_{20} & -b_{14} & -b_{29} & b_{25} & b_{18} & b_{12} & b_{28} & -b_{24} \\
b_{25} & b_{29} & -b_{14} & b_{20} & -b_{24} & -b_{28} & b_{12} & -b_{18} & -b_{23} & -b_{27} & -b_{30} & b_{31} & -b_5 & -b_9 & -b_{15} & b_{21} \\
b_{29} & b_{25} & b_{20} & -b_{14} & -b_{28} & -b_{24} & -b_{18} & b_{12} & -b_{27} & -b_{23} & b_{31} & -b_{30} & -b_9 & -b_5 & b_{21} & -b_{15} \\
-b_{12} & -b_{18} & -b_{24} & b_{28} & b_{14} & b_{20} & b_{25} & -b_{29} & -b_{15} & -b_{21} & b_5 & -b_9 & -b_{30} & -b_{31} & b_{23} & -b_{27} \\
-b_{18} & -b_{12} & b_{28} & -b_{24} & b_{20} & b_{14} & -b_{29} & b_{25} & -b_{21} & -b_{15} & -b_9 & b_5 & -b_{31} & -b_{30} & -b_{27} & b_{23} \\
-b_5 & -b_9 & b_{15} & -b_{21} & -b_{23} & -b_{27} & b_{30} & -b_{31} & b_{24} & b_{28} & b_{12} & -b_{18} & -b_{25} & -b_{29} & -b_{14} & b_{20} \\
-b_9 & -b_5 & -b_{21} & b_{15} & -b_{27} & -b_{23} & -b_{31} & b_{30} & b_{28} & b_{24} & -b_{18} & b_{12} & -b_{29} & -b_{25} & b_{20} & -b_{14} \\
-b_{14} & -b_{20} & -b_{25} & b_{29} & b_{12} & b_{18} & b_{24} & -b_{28} & -b_{30} & -b_{31} & b_{23} & -b_{27} & -b_{15} & -b_{21} & b_5 & -b_9 \\
-b_{20} & -b_{14} & b_{29} & -b_{25} & b_{18} & b_{12} & -b_{28} & b_{24} & -b_{31} & -b_{30} & -b_{27} & b_{23} & -b_{21} & -b_{15} & -b_9 & b_5 \\
b_{23} & b_{27} & -b_{30} & b_{31} & b_5 & b_9 & -b_{15} & b_{21} & -b_{25} & -b_{29} & -b_{14} & b_{20} & b_{24} & b_{28} & b_{12} & -b_{18} \\
b_{27} & b_{23} & b_{31} & -b_{30} & b_9 & b_5 & b_{21} & -b_{15} & -b_{29} & -b_{25} & b_{20} & -b_{14} & b_{28} & b_{24} & -b_{18} & b_{12}
\end{bmatrix}.$$

There is an effective method of factorization of this type matrices, which during the calculation of the matrix-vector products allows to reduce the number of multiplications from 32^2 to $3/4 \cdot 32^2$ at the expense of increasing additions from $32 \cdot 31$ to $5/4 \cdot 32 \cdot 31$ [27]. The matrix $\breve{\mathbf{B}}_{32}$ used in the procedure of multiplication (2) can be described as:

$$\breve{\mathbf{B}}_{32} = \begin{bmatrix} \mathbf{I}_{16} & \mathbf{0}_{16} & \mathbf{I}_{16} \\ \mathbf{0}_{16} & \mathbf{I}_{16} & \mathbf{I}_{16} \end{bmatrix} \begin{bmatrix} (\breve{\mathbf{B}}_{16}^{(0)} - \breve{\mathbf{B}}_{16}^{(1)}) & \mathbf{0}_{16} & \mathbf{0}_{16} \\ \mathbf{0}_{16} & -(\breve{\mathbf{B}}_{16}^{(0)} + \breve{\mathbf{B}}_{16}^{(1)}) & \mathbf{0}_{16} \\ \mathbf{0}_{16} & \mathbf{0}_{16} & \breve{\mathbf{B}}_{16}^{(1)} \end{bmatrix} \begin{bmatrix} \mathbf{I}_{16} & \mathbf{0}_{16} \\ \mathbf{0}_{16} & \mathbf{I}_{16} \\ \mathbf{I}_{16} & \mathbf{I}_{16} \end{bmatrix},$$

where $\mathbf{I}_{16}$ is an identity matrix and $\mathbf{0}_{16}$ is a null matrix. Thus, we can write a new procedure for calculating the product of Kaluza numbers in the following form:

$$\mathbf{Y}_{32\times 1} = \mathbf{M}_{32}^{(r)} \mathbf{T}_{32\times 48} \mathbf{B}_{48} \mathbf{T}_{48\times 32} \mathbf{M}_{32}^{(c)} \mathbf{X}_{32\times 1}, \quad (3)$$

where

$$\mathbf{T}_{32\times 48} = \begin{bmatrix} 1 & 0 & 1 \\ 0 & 1 & 1 \end{bmatrix} \otimes \mathbf{I}_{16},$$

$$\mathbf{B}_{48} = \text{quasidiag}\begin{pmatrix} \mathbf{B}_{16}^{(-)} \\ \mathbf{B}_{16}^{(+)} \\ \mathbf{B}_{16}^{(1)} \end{pmatrix},$$

$$\mathbf{B}_{\mathbf{16}}^{(-)} = \mathbf{B}_{16}^{(0)} - \mathbf{B}_{16}^{(1)}, \tag{4}$$

$$\mathbf{B}_{\mathbf{16}}^{(+)} = -\left(\mathbf{B}_{16}^{(0)} + \mathbf{B}_{16}^{(1)}\right), \tag{5}$$

$$\mathbf{T}_{48\times32} = \begin{bmatrix} 1 & 0 \\ 0 & 1 \\ 1 & 1 \end{bmatrix} \otimes \mathbf{I}_{16},$$

where the symbol "$\otimes$" denotes the tensor product of two matrices and quasidiag() means a block-diagonal matrix.

We introduce the following notation to (4) and (5):

$$\begin{array}{llll}
c_0 = b_{13} + b_{15}, & c_1 = b_{19} + b_{21}, & c_2 = b_5 - b_3, & c_3 = b_7 - b_9, \\
c_4 = b_{30} - b_{11}, & c_5 = b_{31} - b_{17}, & c_6 = b_2 + b_{23}, & c_7 = b_6 + b_{27}, \\
c_8 = b_{12} - b_{10}, & c_9 = b_{18} - b_{16}, & c_{10} = b_{22} + b_{24}, & c_{11} = b_{26} + b_{28}, \\
c_{12} = b_0 + b_{14}, & c_{13} = b_1 + b_{20}, & c_{14} = b_{25} - b_4, & c_{15} = b_8 - b_{29}, \\
c_{16} = b_{11} + b_{30}, & c_{17} = b_{17} + b_{31}, & c_{18} = b_{23} - b_2, & c_{19} = b_6 - b_{27}, \\
c_{20} = b_{15} - b_{13}, & c_{21} = b_{21} - b_{19}, & c_{22} = b_3 + b_5, & c_{23} = b_7 + b_9, \\
c_{24} = b_{14} - b_0, & c_{25} = b_{20} - b_1, & c_{26} = b_4 + b_{25}, & c_{27} = b_8 + b_{29}, \\
c_{28} = b_{10} + b_{12}, & c_{29} = b_{16} + b_{18}, & c_{30} = b_{24} - b_{22}, & c_{31} = b_{26} - b_{28},
\end{array}$$

we obtain:

$$\mathbf{B}_{\mathbf{16}}^{(-)} = \begin{bmatrix}
c_0 & c_1 & c_2 & c_3 & c_4 & c_5 & c_6 & c_7 & c_8 & c_9 & c_{10} & c_{11} & c_{12} & c_{13} & c_{14} & c_{15} \\
c_1 & c_0 & c_3 & c_2 & c_5 & c_4 & c_7 & c_6 & c_9 & c_8 & c_{11} & c_{10} & c_{13} & c_{12} & c_{15} & c_{14} \\
-c_{10} & c_{11} & c_8 & c_9 & c_{14} & c_{15} & c_{12} & c_{13} & c_2 & c_3 & c_0 & c_1 & c_6 & c_7 & c_4 & c_5 \\
-c_{11} & c_{10} & c_9 & c_8 & c_{15} & c_{14} & c_{13} & c_{12} & c_3 & c_2 & c_1 & c_0 & c_7 & c_6 & c_5 & c_4 \\
c_{16} & c_{17} & c_{18} & c_{19} & c_{20} & c_{21} & c_{22} & c_{23} & c_{24} & c_{25} & c_{26} & c_{27} & c_{28} & c_{29} & c_{30} & c_{31} \\
c_{17} & c_{16} & c_{19} & c_{18} & c_{21} & c_{20} & c_{23} & c_{22} & c_{25} & c_{24} & c_{27} & c_{26} & c_{29} & c_{28} & c_{31} & c_{30} \\
c_{26} & c_{27} & c_{24} & c_{25} & c_{30} & c_{31} & c_{28} & c_{29} & c_{18} & c_{19} & c_{16} & c_{17} & c_{22} & c_{23} & c_{20} & c_{21} \\
c_{27} & c_{26} & c_{25} & c_{24} & c_{31} & c_{30} & c_{29} & c_{28} & c_{19} & c_{18} & c_{17} & c_{16} & c_{23} & c_{22} & c_{21} & c_{20} \\
c_8 & c_9 & c_{10} & c_{11} & c_{12} & c_{13} & c_{14} & c_{15} & c_0 & c_1 & c_2 & c_3 & c_4 & c_5 & c_6 & c_7 \\
c_9 & c_8 & c_{11} & c_{10} & c_{13} & c_{12} & c_{15} & c_{14} & c_1 & c_0 & c_3 & c_2 & c_5 & c_4 & c_7 & c_6 \\
c_2 & c_3 & c_0 & c_1 & c_6 & c_7 & c_4 & c_5 & c_{10} & c_{11} & c_8 & c_9 & c_{14} & c_{15} & c_{12} & c_{13} \\
-c_3 & c_2 & c_1 & c_0 & c_7 & c_6 & c_5 & c_4 & c_{11} & c_{10} & c_9 & c_8 & c_{15} & c_{14} & c_{13} & c_{12} \\
c_{24} & c_{25} & c_{26} & c_{27} & c_{28} & c_{29} & c_{30} & c_{31} & c_{16} & c_{17} & c_{18} & c_{19} & c_{20} & c_{21} & c_{22} & c_{23} \\
c_{25} & c_{24} & c_{27} & c_{26} & c_{29} & c_{28} & c_{31} & c_{30} & c_{17} & c_{16} & c_{19} & c_{18} & c_{21} & c_{20} & c_{23} & c_{22} \\
-c_{18} & c_{19} & c_{16} & c_{17} & c_{22} & c_{23} & c_{20} & c_{21} & c_{26} & c_{27} & c_{24} & c_{25} & c_{30} & c_{31} & c_{28} & c_{29} \\
c_{19} & c_{18} & c_{17} & c_{16} & c_{23} & c_{22} & c_{21} & c_{20} & c_{27} & c_{26} & c_{25} & c_{24} & c_{31} & c_{30} & c_{29} & c_{28}
\end{bmatrix},$$

$$\mathbf{B}_{\mathbf{16}}^{(+)} = \begin{bmatrix}
c_{20} & c_{21} & c_{22} & c_{23} & c_{16} & c_{17} & c_{18} & c_{19} & c_{28} & c_{29} & c_{30} & c_{31} & c_{24} & c_{25} & c_{26} & c_{27} \\
c_{21} & c_{20} & c_{23} & c_{22} & c_{17} & c_{16} & c_{19} & c_{18} & c_{29} & c_{28} & c_{31} & c_{30} & c_{25} & c_{24} & c_{27} & c_{26} \\
-c_{30} & c_{31} & c_{28} & c_{29} & c_{26} & c_{27} & c_{24} & c_{25} & c_{22} & c_{23} & c_{20} & c_{21} & c_{18} & c_{19} & c_{16} & c_{17} \\
c_{31} & c_{30} & c_{29} & c_{28} & c_{27} & c_{26} & c_{25} & c_{24} & c_{23} & c_{22} & c_{21} & c_{20} & c_{19} & c_{18} & c_{17} & c_{16} \\
c_4 & c_5 & c_6 & c_7 & c_0 & c_1 & c_2 & c_3 & c_{12} & c_{13} & c_{14} & c_{15} & c_8 & c_9 & c_{10} & c_{11} \\
c_5 & c_4 & c_7 & c_6 & c_1 & c_0 & c_3 & c_2 & c_{13} & c_{12} & c_{15} & c_{14} & c_9 & c_8 & c_{11} & c_{10} \\
c_{14} & c_{15} & c_{12} & c_{13} & c_{10} & c_{11} & c_8 & c_9 & c_6 & c_7 & c_4 & c_5 & c_2 & c_3 & c_0 & c_1 \\
-c_{15} & c_{14} & c_{13} & c_{12} & c_{11} & c_{10} & c_9 & c_8 & c_7 & c_6 & c_5 & c_4 & c_3 & c_2 & c_1 & c_0 \\
c_{28} & c_{29} & c_{30} & c_{31} & c_{24} & c_{25} & c_{26} & c_{27} & c_{20} & c_{21} & c_{22} & c_{23} & c_{16} & c_{17} & c_{18} & c_{19} \\
c_{29} & c_{28} & c_{31} & c_{30} & c_{25} & c_{24} & c_{27} & c_{26} & c_{21} & c_{20} & c_{23} & c_{22} & c_{17} & c_{16} & c_{19} & c_{18} \\
c_{22} & c_{23} & c_{20} & c_{21} & c_{18} & c_{19} & c_{16} & c_{17} & c_{30} & c_{31} & c_{28} & c_{29} & c_{26} & c_{27} & c_{24} & c_{25} \\
c_{23} & c_{22} & c_{21} & c_{20} & c_{19} & c_{18} & c_{17} & c_{16} & c_{31} & c_{30} & c_{29} & c_{28} & c_{27} & c_{26} & c_{25} & c_{24} \\
c_{12} & c_{13} & c_{14} & c_{15} & c_8 & c_9 & c_{10} & c_{11} & c_4 & c_5 & c_6 & c_7 & c_0 & c_1 & c_2 & c_3 \\
c_{13} & c_{12} & c_{15} & c_{14} & c_9 & c_8 & c_{11} & c_{10} & c_5 & c_4 & c_7 & c_6 & c_1 & c_0 & c_3 & c_2 \\
-c_6 & c_7 & c_4 & c_5 & c_2 & c_3 & c_0 & c_1 & c_{14} & c_{15} & c_{12} & c_{13} & c_{10} & c_{11} & c_8 & c_9 \\
-c_7 & c_6 & c_5 & c_4 & c_3 & c_2 & c_1 & c_0 & c_{15} & c_{14} & c_{13} & c_{12} & c_{11} & c_{10} & c_9 & c_8
\end{bmatrix}.$$

The matrices $\mathbf{B}_{16}^{(-)}$, $\mathbf{B}_{16}^{(+)}$ and $\mathbf{B}_{16}^{(1)}$ have similar structures. If we now change the signs of all of the elements of the sixth and seventh rows, as well as all of the elements of the second, third, sixth and seventh columns, to the opposite, then the matrices $\mathbf{B}_{16}^{(-)}$, $\mathbf{B}_{16}^{(+)}$ and $\mathbf{B}_{16}^{(1)}$ will have structures of type $\begin{bmatrix} \mathbf{A}_{N/2} & \mathbf{B}_{N/2} \\ \mathbf{B}_{N/2} & \mathbf{A}_{N/2} \end{bmatrix}$, which leads to reducing the number of real multiplications during matrix-vector product calculation. We can write the sign transformation matrices for rows $\mathbf{S}_{16}^{(r)}$ and columns $\mathbf{S}_{16}^{(c)}$ as:

$$\mathbf{S}_{16}^{(r)} = \mathrm{diag}(1,\ 1,\ 1,\ 1,\ 1,\ 1,\ -1,\ -1,\ 1,\ 1,\ 1,\ 1,\ 1,\ 1,\ 1,\ 1),$$

$$\mathbf{S}_{16}^{(c)} = \mathrm{diag}(1,\ 1,\ -1,\ -1,\ 1,\ 1,\ -1,\ -1,\ 1,\ 1,\ 1,\ 1,\ 1,\ 1,\ 1,\ 1).$$

Then, we obtain new standardized matrices:

$$\breve{\mathbf{B}}_{16}^{(-)} = \mathbf{S}_{16}^{(r)} \mathbf{B}_{16}^{(-)} \mathbf{S}_{16}^{(c)} = \begin{bmatrix} \mathbf{B}_8^{(0)} & \mathbf{B}_8^{(1)} \\ \mathbf{B}_8^{(1)} & \mathbf{B}_8^{(0)} \end{bmatrix}, \tag{6}$$

$$\breve{\mathbf{B}}_{16}^{(+)} = \mathbf{S}_{16}^{(r)} \mathbf{B}_{16}^{(+)} \mathbf{S}_{16}^{(c)} = \begin{bmatrix} \mathbf{B}_8^{(2)} & \mathbf{B}_8^{(3)} \\ \mathbf{B}_8^{(3)} & \mathbf{B}_8^{(2)} \end{bmatrix}, \tag{7}$$

$$\breve{\mathbf{B}}_{16}^{(1)} = \mathbf{S}_{16}^{(r)} \mathbf{B}_{16}^{(1)} \mathbf{S}_{16}^{(c)} = \begin{bmatrix} \mathbf{B}_8^{(4)} & \mathbf{B}_8^{(5)} \\ \mathbf{B}_8^{(5)} & \mathbf{B}_8^{(4)} \end{bmatrix}, \tag{8}$$

where:

$$\mathbf{B}_8^{(0)} = \begin{bmatrix} c_0 & c_1 & c_2 & c_3 & c_4 & c_5 & c_6 & c_7 \\ c_1 & c_0 & c_3 & c_2 & c_5 & c_4 & c_7 & c_6 \\ -c_{10} & c_{11} & c_8 & c_9 & c_{14} & c_{15} & c_{12} & c_{13} \\ -c_{11} & c_{10} & c_9 & c_8 & c_{15} & c_{14} & c_{13} & c_{12} \\ c_{16} & c_{17} & c_{18} & c_{19} & c_{20} & c_{21} & c_{22} & c_{23} \\ c_{17} & c_{16} & c_{19} & c_{18} & c_{21} & c_{20} & c_{23} & c_{22} \\ c_{26} & c_{27} & c_{24} & c_{25} & c_{30} & c_{31} & c_{28} & c_{29} \\ c_{27} & c_{26} & c_{25} & c_{24} & c_{31} & c_{30} & c_{29} & c_{28} \end{bmatrix},$$

$$\mathbf{B}_8^{(1)} = \begin{bmatrix} c_8 & c_9 & c_{10} & c_{11} & c_{12} & c_{13} & c_{14} & c_{15} \\ c_9 & c_8 & c_{11} & c_{10} & c_{13} & c_{12} & c_{15} & c_{14} \\ c_2 & c_3 & c_0 & c_1 & c_6 & c_7 & c_4 & c_5 \\ -c_3 & c_2 & c_1 & c_0 & c_7 & c_6 & c_5 & c_4 \\ c_{24} & c_{25} & c_{26} & c_{27} & c_{28} & c_{29} & c_{30} & c_{31} \\ c_{25} & c_{24} & c_{27} & c_{26} & c_{29} & c_{28} & c_{31} & c_{30} \\ -c_{18} & c_{19} & c_{16} & c_{17} & c_{22} & c_{23} & c_{20} & c_{21} \\ c_{19} & c_{18} & c_{17} & c_{16} & c_{23} & c_{22} & c_{21} & c_{20} \end{bmatrix},$$

$$\mathbf{B}_8^{(2)} = \begin{bmatrix} c_{20} & c_{21} & c_{22} & c_{23} & c_{16} & c_{17} & c_{18} & c_{19} \\ c_{21} & c_{20} & c_{23} & c_{22} & c_{17} & c_{16} & c_{19} & c_{18} \\ -c_{30} & c_{31} & c_{28} & c_{29} & c_{26} & c_{27} & c_{24} & c_{25} \\ c_{31} & c_{30} & c_{29} & c_{28} & c_{27} & c_{26} & c_{25} & c_{24} \\ c_4 & c_5 & c_6 & c_7 & c_0 & c_1 & c_2 & c_3 \\ c_5 & c_4 & c_7 & c_6 & c_1 & c_0 & c_3 & c_2 \\ c_{14} & c_{15} & c_{12} & c_{13} & c_{10} & c_{11} & c_8 & c_9 \\ -c_{15} & c_{14} & c_{13} & c_{12} & c_{11} & c_{10} & c_9 & c_8 \end{bmatrix},$$

$$\mathbf{B}_8^{(3)} = \begin{bmatrix} c_{28} & c_{29} & c_{30} & c_{31} & c_{24} & c_{25} & c_{26} & c_{27} \\ c_{29} & c_{28} & c_{31} & c_{30} & c_{25} & c_{24} & c_{27} & c_{26} \\ c_{22} & c_{23} & c_{20} & c_{21} & c_{18} & c_{19} & c_{16} & c_{17} \\ c_{23} & c_{22} & c_{21} & c_{20} & c_{19} & c_{18} & c_{17} & c_{16} \\ c_{12} & c_{13} & c_{14} & c_{15} & c_{8} & c_{9} & c_{10} & c_{11} \\ c_{13} & c_{12} & c_{15} & c_{14} & c_{9} & c_{8} & c_{11} & c_{10} \\ -c_{6} & c_{7} & c_{4} & c_{5} & c_{2} & c_{3} & c_{0} & c_{1} \\ -c_{7} & c_{6} & c_{5} & c_{4} & c_{3} & c_{2} & c_{1} & c_{0} \end{bmatrix},$$

$$\mathbf{B}_8^{(4)} = \begin{bmatrix} -b_{15} & -b_{21} & b_{5} & -b_{9} & -b_{30} & -b_{31} & b_{23} & -b_{27} \\ -b_{21} & -b_{15} & -b_{9} & b_{5} & -b_{31} & -b_{30} & -b_{27} & b_{23} \\ b_{24} & b_{28} & b_{12} & -b_{18} & -b_{25} & -b_{29} & -b_{14} & b_{20} \\ b_{28} & b_{24} & -b_{18} & b_{12} & -b_{29} & -b_{25} & b_{20} & -b_{14} \\ -b_{30} & -b_{31} & b_{23} & -b_{27} & -b_{15} & -b_{21} & b_{5} & -b_{9} \\ -b_{31} & -b_{30} & -b_{27} & b_{23} & -b_{21} & -b_{15} & -b_{9} & b_{5} \\ -b_{25} & -b_{29} & -b_{14} & b_{20} & b_{24} & b_{28} & b_{12} & -b_{18} \\ -b_{29} & -b_{25} & b_{20} & -b_{14} & b_{28} & b_{24} & -b_{18} & b_{12} \end{bmatrix},$$

$$\mathbf{B}_8^{(5)} = \begin{bmatrix} -b_{12} & -b_{18} & b_{24} & -b_{28} & b_{14} & b_{20} & -b_{25} & b_{29} \\ -b_{18} & -b_{12} & -b_{28} & b_{24} & b_{20} & b_{14} & b_{29} & -b_{25} \\ -b_{5} & -b_{9} & -b_{15} & b_{21} & -b_{23} & -b_{27} & -b_{30} & b_{31} \\ -b_{9} & -b_{5} & b_{21} & -b_{15} & -b_{27} & -b_{23} & b_{31} & -b_{30} \\ -b_{14} & -b_{20} & b_{25} & -b_{29} & b_{12} & b_{18} & -b_{24} & b_{28} \\ -b_{20} & -b_{14} & -b_{29} & b_{25} & b_{18} & b_{12} & b_{28} & -b_{24} \\ b_{23} & b_{27} & b_{30} & -b_{31} & b_{5} & b_{9} & b_{15} & -b_{21} \\ b_{27} & b_{23} & -b_{31} & b_{30} & b_{9} & b_{5} & -b_{21} & b_{15} \end{bmatrix}.$$

There is a possiblity to use a method of factorization for the standardized matrices (6)–(8). This allows us to reduce the number of multiplications to $8^2/2$ using $8(8+1)$ additions for each of above matrices. Therefore, similar to the previous we can write [27,28]:

$$\begin{bmatrix} \mathbf{A}_{N/2} & \mathbf{B}_{N/2} \\ \mathbf{B}_{N/2} & \mathbf{A}_{N/2} \end{bmatrix} = \begin{bmatrix} \mathbf{I}_{N/2} & \mathbf{I}_{N/2} \\ \mathbf{I}_{N/2} & -\mathbf{I}_{N/2} \end{bmatrix} \begin{bmatrix} \frac{1}{2}(\mathbf{A}_{N/2} + \mathbf{B}_{N/2}) & \mathbf{0}_{N/2} \\ \mathbf{0}_{N/2} & \frac{1}{2}(\mathbf{A}_{N/2} - \mathbf{B}_{N/2}) \end{bmatrix} \begin{bmatrix} \mathbf{I}_{N/2} & \mathbf{I}_{N/2} \\ \mathbf{I}_{N/2} & -\mathbf{I}_{N/2} \end{bmatrix}, \quad (9)$$

where $\mathbf{A}_{N/2}$, $\mathbf{B}_{N/2}$ are some matrices. Therefore, we can rewrite (6)–(8) as:

$$\breve{\mathbf{B}}_{16}^{(-)} = \begin{bmatrix} \mathbf{I}_8 & \mathbf{I}_8 \\ \mathbf{I}_8 & -\mathbf{I}_8 \end{bmatrix} \begin{bmatrix} \frac{1}{2}\mathbf{B}_8^{(0+)} & \mathbf{0}_8 \\ \mathbf{0}_8 & \frac{1}{2}\mathbf{B}_8^{(0-)} \end{bmatrix} \begin{bmatrix} \mathbf{I}_8 & \mathbf{I}_8 \\ \mathbf{I}_8 & -\mathbf{I}_8 \end{bmatrix},$$

$$\breve{\mathbf{B}}_{16}^{(+)} = \begin{bmatrix} \mathbf{I}_8 & \mathbf{I}_8 \\ \mathbf{I}_8 & -\mathbf{I}_8 \end{bmatrix} \begin{bmatrix} \frac{1}{2}\mathbf{B}_8^{(1+)} & \mathbf{0}_8 \\ \mathbf{0}_8 & \frac{1}{2}\mathbf{B}_8^{(1-)} \end{bmatrix} \begin{bmatrix} \mathbf{I}_8 & \mathbf{I}_8 \\ \mathbf{I}_8 & -\mathbf{I}_8 \end{bmatrix},$$

$$\breve{\mathbf{B}}_{16}^{(1)} = \begin{bmatrix} \mathbf{I}_8 & \mathbf{I}_8 \\ \mathbf{I}_8 & -\mathbf{I}_8 \end{bmatrix} \begin{bmatrix} \frac{1}{2}\mathbf{B}_8^{(2+)} & \mathbf{0}_8 \\ \mathbf{0}_8 & \frac{1}{2}\mathbf{B}_8^{(2-)} \end{bmatrix} \begin{bmatrix} \mathbf{I}_8 & \mathbf{I}_8 \\ \mathbf{I}_8 & -\mathbf{I}_8 \end{bmatrix},$$

where:

$$\mathbf{B}_8^{(0+)} = \mathbf{B}_8^{(0)} + \mathbf{B}_8^{(1)}, \quad (10)$$

$$\mathbf{B}_8^{(0-)} = \mathbf{B}_8^{(0)} - \mathbf{B}_8^{(1)}, \quad (11)$$

$$\mathbf{B}_8^{(1+)} = \mathbf{B}_8^{(2)} + \mathbf{B}_8^{(3)}, \quad (12)$$

$$\mathbf{B}_8^{(1-)} = \mathbf{B}_8^{(2)} - \mathbf{B}_8^{(3)}, \quad (13)$$

$$\mathbf{B}_8^{(2+)} = \mathbf{B}_8^{(4)} + \mathbf{B}_8^{(5)}, \quad (14)$$

$$\mathbf{B}_8^{(2-)} = \mathbf{B}_8^{(4)} - \mathbf{B}_8^{(5)}. \quad (15)$$

Combining partial decompositions in a single procedure we can rewrite procedure, (3) as following:

$$\mathbf{Y}_{32\times 1} = \mathbf{M}_{32}^{(r)}\mathbf{T}_{32\times 48}\mathbf{S}_{48}^{(r)}\mathbf{W}_{48}^{(1)}\tilde{\mathbf{B}}_{48}\mathbf{W}_{48}^{(1)}\mathbf{S}_{48}^{(c)}\mathbf{T}_{48\times 32}\mathbf{M}_{32}^{(c)}\mathbf{X}_{32\times 1},$$

where

$$\tilde{\mathbf{B}}_{48} = \text{quasidiag}\left(\frac{1}{2}\mathbf{B}_8^{(0+)}, \frac{1}{2}\mathbf{B}_8^{(0-)}, \frac{1}{2}\mathbf{B}_8^{(1+)}, \frac{1}{2}\mathbf{B}_8^{(1-)}, \frac{1}{2}\mathbf{B}_8^{(2+)}, \frac{1}{2}\mathbf{B}_8^{(2-)}\right),$$

$$\mathbf{S}_{48}^{(r)} = \mathbf{I}_3 \otimes \mathbf{S}_{16}^{(r)},$$

$$\mathbf{S}_{48}^{(c)} = \mathbf{I}_3 \otimes \mathbf{S}_{16}^{(c)},$$

$$\mathbf{W}_{48}^{(1)} = \mathbf{I}_3 \otimes \mathbf{H}_2 \otimes \mathbf{I}_8,$$

$\mathbf{H}_2$ is the order 2 Hadamard matrix, i.e.:

$$\mathbf{H}_2 = \begin{bmatrix} 1 & 1 \\ 1 & -1 \end{bmatrix}.$$

Introducing the following notation:

$$\begin{array}{llll}
d_0 = c_0 + c_8, & d_1 = c_1 + c_9, & d_2 = c_2 + c_{10}, & d_3 = c_{11} - c_3, \\
d_4 = c_4 - c_{12}, & d_5 = c_5 - c_{13}, & d_6 = c_{14} - c_6, & d_7 = c_7 + c_{15}, \\
d_8 = c_2 - c_{10}, & d_9 = c_3 + c_{11}, & d_{10} = c_0 - c_8, & d_{11} = c_9 - c_1, \\
d_{12} = c_6 + c_{14}, & d_{13} = c_7 - c_{15}, & d_{14} = c_4 + c_{12}, & d_{15} = c_5 + c_{13}, \\
d_{16} = c_{16} + c_{24}, & d_{17} = c_{17} + c_{25}, & d_{18} = c_{18} + c_{26}, & d_{19} = c_{27} - c_{19}, \\
d_{20} = c_{20} - c_{28}, & d_{21} = c_{21} - c_{29}, & d_{22} = c_{30} - c_{22}, & d_{23} = c_{23} + c_{31}, \\
d_{24} = c_{26} - c_{18}, & d_{25} = c_{19} + c_{27}, & d_{26} = c_{24} - c_{16}, & d_{27} = c_{17} - c_{25}, \\
d_{28} = c_{22} + c_{30}, & d_{29} = c_{31} - c_{23}, & d_{30} = c_{20} + c_{28}, & d_{31} = c_{21} + c_{29}.
\end{array}$$

to (10)–(13), we obtain:

$$\mathbf{B}_8^{(0+)} = \begin{bmatrix}
d_0 & d_1 & d_2 & d_3 & d_4 & d_5 & d_6 & d_7 \\
d_1 & d_0 & d_3 & d_2 & d_5 & d_4 & d_7 & d_6 \\
d_8 & d_9 & d_{10} & d_{11} & d_{12} & d_{13} & d_{14} & d_{15} \\
-d_9 & d_8 & d_{11} & d_{10} & d_{13} & d_{12} & d_{15} & d_{14} \\
d_{16} & d_{17} & d_{18} & d_{19} & d_{20} & d_{21} & d_{22} & d_{23} \\
d_{17} & d_{16} & d_{19} & d_{18} & d_{21} & d_{20} & d_{23} & d_{22} \\
d_{24} & d_{25} & d_{26} & d_{27} & d_{28} & d_{29} & d_{30} & d_{31} \\
d_{25} & d_{24} & d_{27} & d_{26} & d_{29} & d_{28} & d_{31} & d_{30}
\end{bmatrix},$$

$$\mathbf{B}_8^{(0-)} = \begin{bmatrix}
d_{10} & d_{11} & d_8 & d_9 & d_{14} & d_{15} & d_{12} & d_{13} \\
-d_{11} & d_{10} & d_9 & d_8 & d_{15} & d_{14} & d_{13} & d_{12} \\
-d_2 & d_3 & d_0 & d_1 & d_6 & d_7 & d_4 & d_5 \\
-d_3 & d_2 & d_1 & d_0 & d_7 & d_6 & d_5 & d_4 \\
-d_{26} & d_{27} & d_{24} & d_{25} & d_{30} & d_{31} & d_{28} & d_{29} \\
d_{27} & d_{26} & d_{25} & d_{24} & d_{31} & d_{30} & d_{29} & d_{28} \\
d_{18} & d_{19} & d_{16} & d_{17} & d_{22} & d_{23} & d_{20} & d_{21} \\
d_{19} & d_{18} & d_{17} & d_{16} & d_{23} & d_{22} & d_{21} & d_{20}
\end{bmatrix},$$

$$\mathbf{B}_8^{(1+)} = \begin{bmatrix} d_{30} & d_{31} & d_{28} & d_{29} & d_{26} & d_{27} & d_{24} & d_{25} \\ d_{31} & d_{30} & d_{29} & d_{28} & d_{27} & d_{26} & d_{25} & d_{24} \\ -d_{22} & d_{23} & d_{20} & d_{21} & d_{18} & d_{19} & d_{16} & d_{17} \\ d_{23} & d_{22} & d_{21} & d_{20} & d_{19} & d_{18} & d_{17} & d_{16} \\ d_{14} & d_{15} & d_{12} & d_{13} & d_{10} & d_{11} & d_{8} & d_{9} \\ d_{15} & d_{14} & d_{13} & d_{12} & d_{11} & d_{10} & d_{9} & d_{8} \\ d_{6} & d_{7} & d_{4} & d_{5} & d_{2} & d_{3} & d_{0} & d_{1} \\ -d_{7} & d_{6} & d_{5} & d_{4} & d_{3} & d_{2} & d_{1} & d_{0} \end{bmatrix},$$

$$\mathbf{B}_8^{(1-)} = \begin{bmatrix} d_{20} & d_{21} & d_{22} & d_{23} & d_{16} & d_{17} & d_{18} & d_{19} \\ d_{21} & d_{20} & d_{23} & d_{22} & d_{17} & d_{16} & d_{19} & d_{18} \\ -d_{28} & d_{29} & d_{30} & d_{31} & d_{24} & d_{25} & d_{26} & d_{27} \\ d_{29} & d_{28} & d_{31} & d_{30} & d_{25} & d_{24} & d_{27} & d_{26} \\ d_{4} & d_{5} & d_{6} & d_{7} & d_{0} & d_{1} & d_{2} & d_{3} \\ d_{5} & d_{4} & d_{7} & d_{6} & d_{1} & d_{0} & d_{3} & d_{2} \\ d_{12} & d_{13} & d_{14} & d_{15} & d_{8} & d_{9} & d_{10} & d_{11} \\ d_{13} & d_{12} & d_{15} & d_{14} & d_{9} & d_{8} & d_{11} & d_{10} \end{bmatrix}.$$

In order to simplify, we introduce the following notation for the elements of matrix $\mathbf{B}_8^{(2+)}$ (14):

$$\begin{array}{llll} c_{32} = b_{12} + b_{15}, & c_{33} = b_{18} + b_{21}, & c_{34} = b5 + b_{24}, & c_{35} = b9 + b_{28}, \\ c_{36} = b_{14} - b_{30}, & c_{37} = b_{20} - b_{31}, & c_{38} = b_{23} - b_{25}, & c_{39} = b_{29} - b_{27}, \\ c_{40} = b_{24} - b5, & c_{41} = b_{28} - b9, & c_{42} = b_{12} - b_{15}, & c_{43} = b_{21} - b_{18}, \\ c_{44} = b_{23} + b_{25}, & c_{45} = b_{27} + b_{29}, & c_{46} = b_{14} + b_{30}, & c_{47} = b_{20} + b_{31}, \end{array}$$

we obtain:

$$\mathbf{B}_8^{(2+)} = \begin{bmatrix} -c_{32} & c_{33} & c_{34} & c_{35} & c_{36} & c_{37} & c_{38} & c_{39} \\ -c_{33} & c_{32} & c_{35} & c_{34} & c_{37} & c_{36} & c_{39} & c_{38} \\ c_{40} & c_{41} & c_{42} & c_{43} & c_{44} & c_{45} & c_{46} & c_{47} \\ c_{41} & c_{40} & c_{43} & c_{42} & c_{45} & c_{44} & c_{47} & c_{46} \\ -c_{46} & c_{47} & c_{44} & c_{45} & c_{42} & c_{43} & c_{40} & c_{41} \\ -c_{47} & c_{46} & c_{45} & c_{44} & c_{43} & c_{42} & c_{41} & c_{40} \\ c_{38} & c_{39} & c_{36} & c_{37} & c_{34} & c_{35} & c_{32} & c_{33} \\ -c_{39} & c_{38} & c_{37} & c_{36} & c_{35} & c_{34} & c_{33} & c_{32} \end{bmatrix}.$$

Now, we introduce the following notation for the elements of matrix $\mathbf{B}_8^{(2-)}$ (15):

$$\begin{array}{llll} c_{48} = b_{12} - b_{15}, & c_{49} = b_{18} - b_{21}, & c_{50} = b5 - b_{24}, & c_{51} = b_{28} - b9, \\ c_{52} = b_{14} + b_{30}, & c_{53} = b_{20} + b_{31}, & c_{54} = b_{23} + b_{25}, & c_{55} = b_{27} + b_{29}, \\ c_{56} = b5 + b_{24}, & c_{57} = b9 + b_{28}, & c_{58} = b_{12} + b_{15}, & c_{59} = b_{18} + b_{21}, \\ c_{60} = b_{23} - b_{25}, & c_{61} = b_{27} - b_{29}, & c_{62} = b_{30} - b_{14}, & c_{63} = b_{20} - b_{31}, \end{array}$$

we obtain:

$$\mathbf{B}_8^{(2-)} = \begin{bmatrix} c_{48} & c_{49} & c_{50} & c_{51} & c_{52} & c_{53} & c_{54} & c_{55} \\ c_{49} & c_{48} & c_{51} & c_{50} & c_{53} & c_{52} & c_{55} & c_{54} \\ c_{56} & c_{57} & c_{58} & c_{59} & c_{60} & c_{61} & c_{62} & c_{63} \\ c_{57} & c_{56} & c_{59} & c_{58} & c_{61} & c_{60} & c_{63} & c_{62} \\ -c_{62} & c_{63} & c_{60} & c_{61} & c_{58} & c_{59} & c_{56} & c_{57} \\ c_{63} & c_{62} & c_{61} & c_{60} & c_{59} & c_{58} & c_{57} & c_{56} \\ -c_{54} & c_{55} & c_{52} & c_{53} & c_{50} & c_{51} & c_{48} & c_{49} \\ -c_{55} & c_{54} & c_{53} & c_{52} & c_{51} & c_{50} & c_{49} & c_{48} \end{bmatrix}.$$

All of the above matrices have the same internal structure. We can permute rows and columns using the π_r = (5 1 2 7 4 0 3 6) and π_c = (5 1 2 6 4 0 3 7) permutation rules, respectively. We obtain the following form:

$$\mathbf{B}_8^{(\gamma)} = \mathbf{P}_8^{(r)} \hat{\mathbf{B}}_8^{(x)} \mathbf{P}_8^{(c)}, \tag{16}$$

where $\mathbf{B}_8^{(\gamma)}$, $\hat{\mathbf{B}}_8^{(\gamma)}$ are the corresponding items in the sets:

$$\mathbf{B}_8^{(\gamma)} \in \left\{ \mathbf{B}_8^{(0+)}, \mathbf{B}_8^{(0-)}, \mathbf{B}_8^{(1+)}, \mathbf{B}_8^{(1-)}, \mathbf{B}_8^{(2+)}, \mathbf{B}_8^{(2-)} \right\},$$
$$\hat{\mathbf{B}}_8^{(\gamma)} \in \left\{ \hat{\mathbf{B}}_8^{(0+)}, \hat{\mathbf{B}}_8^{(0-)}, \hat{\mathbf{B}}_8^{(1+)}, \hat{\mathbf{B}}_8^{(1-)}, \hat{\mathbf{B}}_8^{(2+)}, \hat{\mathbf{B}}_8^{(2-)} \right\}$$

and

$$\mathbf{P}_8^{(r)} = \begin{bmatrix} 0 & 0 & 0 & 0 & 0 & 1 & 0 & 0 \\ 0 & 1 & 0 & 0 & 0 & 0 & 0 & 0 \\ 0 & 0 & 1 & 0 & 0 & 0 & 0 & 0 \\ 0 & 0 & 0 & 0 & 0 & 0 & 1 & 0 \\ 0 & 0 & 0 & 0 & 1 & 0 & 0 & 0 \\ 1 & 0 & 0 & 0 & 0 & 0 & 0 & 0 \\ 0 & 0 & 0 & 0 & 0 & 0 & 0 & 1 \\ 0 & 0 & 0 & 1 & 0 & 0 & 0 & 0 \end{bmatrix}, \quad \mathbf{P}_8^{(c)} = \begin{bmatrix} 0 & 0 & 0 & 0 & 0 & 1 & 0 & 0 \\ 0 & 1 & 0 & 0 & 0 & 0 & 0 & 0 \\ 0 & 0 & 1 & 0 & 0 & 0 & 0 & 0 \\ 0 & 0 & 0 & 0 & 0 & 0 & 1 & 0 \\ 0 & 0 & 0 & 0 & 1 & 0 & 0 & 0 \\ 1 & 0 & 0 & 0 & 0 & 0 & 0 & 0 \\ 0 & 0 & 0 & 1 & 0 & 0 & 0 & 0 \\ 0 & 0 & 0 & 0 & 0 & 0 & 0 & 1 \end{bmatrix}.$$

The matrices $\hat{\mathbf{B}}_8^{(\gamma)}$ (16) are calculated via the following equation:

$$\hat{\mathbf{B}}_8^{(\gamma)} = \left(\mathbf{P}_8^{(r)}\right)^{-1} \mathbf{B}_8^{(x)} \left(\mathbf{P}_8^{(c)}\right)^{-1}$$

and have a standardized form (9) that reduces the number of multiplications. Thus, we can write:

$$\hat{\mathbf{B}}_8^{(0+)} = \left(\mathbf{P}_8^{(r)}\right)^{-1} \mathbf{B}_8^{(0+)} \left(\mathbf{P}_8^{(c)}\right)^{-1} = \begin{bmatrix} \mathbf{B}_4^{(0)} & \mathbf{B}_4^{(1)} \\ \mathbf{B}_4^{(1)} & \mathbf{B}_4^{(0)} \end{bmatrix},$$

$$\hat{\mathbf{B}}_8^{(0-)} = \left(\mathbf{P}_8^{(r)}\right)^{-1} \mathbf{B}_8^{(0-)} \left(\mathbf{P}_8^{(c)}\right)^{-1} = \begin{bmatrix} \mathbf{B}_4^{(2)} & \mathbf{B}_4^{(3)} \\ \mathbf{B}_4^{(3)} & \mathbf{B}_4^{(2)} \end{bmatrix},$$

$$\hat{\mathbf{B}}_8^{(1+)} = \left(\mathbf{P}_8^{(r)}\right)^{-1} \mathbf{B}_8^{(1+)} \left(\mathbf{P}_8^{(c)}\right)^{-1} = \begin{bmatrix} \mathbf{B}_4^{(4)} & \mathbf{B}_4^{(5)} \\ \mathbf{B}_4^{(5)} & \mathbf{B}_4^{(4)} \end{bmatrix},$$

$$\hat{\mathbf{B}}_8^{(1-)} = \left(\mathbf{P}_8^{(r)}\right)^{-1} \mathbf{B}_8^{(1-)} \left(\mathbf{P}_8^{(c)}\right)^{-1} = \begin{bmatrix} \mathbf{B}_4^{(6)} & \mathbf{B}_4^{(7)} \\ \mathbf{B}_4^{(7)} & \mathbf{B}_4^{(6)} \end{bmatrix},$$

$$\hat{\mathbf{B}}_8^{(2+)} = \left(\mathbf{P}_8^{(r)}\right)^{-1} \mathbf{B}_8^{(2+)} \left(\mathbf{P}_8^{(c)}\right)^{-1} = \begin{bmatrix} \mathbf{B}_4^{(8)} & \mathbf{B}_4^{(9)} \\ \mathbf{B}_4^{(9)} & \mathbf{B}_4^{(8)} \end{bmatrix},$$

$$\hat{\mathbf{B}}_8^{(2-)} = \left(\mathbf{P}_8^{(r)}\right)^{-1} \mathbf{B}_8^{(2-)} \left(\mathbf{P}_8^{(c)}\right)^{-1} = \begin{bmatrix} \mathbf{B}_4^{(10)} & \mathbf{B}_4^{(11)} \\ \mathbf{B}_4^{(11)} & \mathbf{B}_4^{(10)} \end{bmatrix},$$

where:

$$\mathbf{B}_4^{(0)} = \begin{bmatrix} d_{20} & d_{16} & d_{19} & d_{23} \\ d_4 & d_0 & d_3 & d_7 \\ d_{13} & d_9 & d_{10} & d_{14} \\ -d_{28} & d_{24} & d_{27} & d_{31} \end{bmatrix}, \quad \mathbf{B}_4^{(1)} = \begin{bmatrix} d_{21} & d_{17} & d_{18} & d_{22} \\ d_5 & d_1 & d_2 & d_6 \\ d_{12} & d_8 & d_{11} & d_{15} \\ d_{29} & d_{25} & d_{26} & d_{30} \end{bmatrix},$$

$$\mathbf{B}_4^{(2)} = \begin{bmatrix} d_{30} & d_{26} & d_{25} & d_{29} \\ d_{14} & d_{10} & d_9 & d_{13} \\ -d_7 & d_3 & d_0 & d_4 \\ -d_{22} & d_{18} & d_{17} & d_{21} \end{bmatrix}, \quad \mathbf{B}_4^{(3)} = \begin{bmatrix} d_{31} & d_{27} & d_{24} & d_{28} \\ d_{15} & d_{11} & d_8 & d_{12} \\ d_6 & d_2 & d_1 & d_5 \\ d_{23} & d_{19} & d_{16} & d_{20} \end{bmatrix},$$

$$\mathbf{B}_4^{(4)} = \begin{bmatrix} d_{10} & d_{14} & d_{13} & d_9 \\ -d_{26} & d_{30} & d_{29} & d_{25} \\ d_{19} & d_{23} & d_{20} & d_{16} \\ -d_2 & d_6 & d_5 & d_1 \end{bmatrix}, \quad \mathbf{B}_4^{(5)} = \begin{bmatrix} -d_{11} & d_{15} & d_{12} & d_8 \\ d_{27} & d_{31} & d_{28} & d_{24} \\ d_{18} & d_{22} & d_{21} & d_{17} \\ -d_3 & d_7 & d_4 & d_0 \end{bmatrix},$$

$$\mathbf{B}_4^{(6)} = \begin{bmatrix} d_0 & d_4 & d_7 & d_3 \\ d_{16} & d_{20} & d_{23} & d_{19} \\ d_{25} & d_{29} & d_{30} & d_{26} \\ d_8 & d_{12} & d_{15} & d_{11} \end{bmatrix}, \quad \mathbf{B}_4^{(7)} = \begin{bmatrix} d_1 & d_5 & d_6 & d_2 \\ d_{17} & d_{21} & d_{22} & d_{18} \\ d_{24} & d_{28} & d_{31} & d_{27} \\ -d_9 & d_{13} & d_{14} & d_{10} \end{bmatrix},$$

$$\mathbf{B}_4^{(8)} = \begin{bmatrix} c_{42} & c_{46} & c_{45} & c_{41} \\ c_{36} & c_{32} & c_{35} & c_{39} \\ -c_{45} & c_{41} & c_{42} & c_{46} \\ c_{34} & c_{38} & c_{37} & c_{33} \end{bmatrix}, \quad \mathbf{B}_4^{(9)} = \begin{bmatrix} -c_{43} & c_{47} & c_{44} & c_{40} \\ c_{37} & c_{33} & c_{34} & c_{38} \\ -c_{44} & c_{40} & c_{43} & c_{47} \\ c_{35} & c_{39} & c_{36} & c_{32} \end{bmatrix},$$

$$\mathbf{B}_4^{(10)} = \begin{bmatrix} -c_{58} & c_{62} & c_{61} & c_{57} \\ -c_{52} & c_{48} & c_{51} & c_{55} \\ c_{61} & c_{57} & c_{58} & c_{62} \\ -c_{50} & c_{54} & c_{53} & c_{49} \end{bmatrix}, \quad \mathbf{B}_4^{(11)} = \begin{bmatrix} -c_{59} & c_{63} & c_{60} & c_{56} \\ -c_{53} & c_{49} & c_{50} & c_{54} \\ c_{60} & c_{56} & c_{59} & c_{63} \\ c_{51} & c_{55} & c_{52} & c_{48} \end{bmatrix}.$$

We can use the multiplication procedure (9) and represent the above matrices in a form:

$$\hat{\mathbf{B}}_8^{(0+)} = \begin{bmatrix} \mathbf{I}_4 & \mathbf{I}_4 \\ \mathbf{I}_4 & -\mathbf{I}_4 \end{bmatrix} \begin{bmatrix} \frac{1}{2}\left(\mathbf{B}_4^{(0)} + \mathbf{B}_4^{(1)}\right) & \mathbf{0} \\ \mathbf{0} & \frac{1}{2}\left(\mathbf{B}_4^{(0)} - \mathbf{B}_4^{(1)}\right) \end{bmatrix} \begin{bmatrix} \mathbf{I}_4 & \mathbf{I}_4 \\ \mathbf{I}_4 & -\mathbf{I}_4 \end{bmatrix},$$

$$\hat{\mathbf{B}}_8^{(0-)} = \begin{bmatrix} \mathbf{I}_4 & \mathbf{I}_4 \\ \mathbf{I}_4 & -\mathbf{I}_4 \end{bmatrix} \begin{bmatrix} \frac{1}{2}\left(\mathbf{B}_4^{(2)} + \mathbf{B}_4^{(3)}\right) & \mathbf{0} \\ \mathbf{0} & \frac{1}{2}\left(\mathbf{B}_4^{(2)} - \mathbf{B}_4^{(3)}\right) \end{bmatrix} \begin{bmatrix} \mathbf{I}_4 & \mathbf{I}_4 \\ \mathbf{I}_4 & -\mathbf{I}_4 \end{bmatrix},$$

$$\hat{\mathbf{B}}_8^{(1+)} = \begin{bmatrix} \mathbf{I}_4 & \mathbf{I}_4 \\ \mathbf{I}_4 & -\mathbf{I}_4 \end{bmatrix} \begin{bmatrix} \frac{1}{2}\left(\mathbf{B}_4^{(4)} + \mathbf{B}_4^{(5)}\right) & \mathbf{0} \\ \mathbf{0} & \frac{1}{2}\left(\mathbf{B}_4^{(4)} - \mathbf{B}_4^{(5)}\right) \end{bmatrix} \begin{bmatrix} \mathbf{I}_4 & \mathbf{I}_4 \\ \mathbf{I}_4 & -\mathbf{I}_4 \end{bmatrix},$$

$$\hat{\mathbf{B}}_8^{(1-)} = \begin{bmatrix} \mathbf{I}_4 & \mathbf{I}_4 \\ \mathbf{I}_4 & -\mathbf{I}_4 \end{bmatrix} \begin{bmatrix} \frac{1}{2}\left(\mathbf{B}_4^{(6)} + \mathbf{B}_4^{(7)}\right) & \mathbf{0} \\ \mathbf{0} & \frac{1}{2}\left(\mathbf{B}_4^{(6)} - \mathbf{B}_4^{(7)}\right) \end{bmatrix} \begin{bmatrix} \mathbf{I}_4 & \mathbf{I}_4 \\ \mathbf{I}_4 & -\mathbf{I}_4 \end{bmatrix},$$

$$\hat{\mathbf{B}}_8^{(2+)} = \begin{bmatrix} \mathbf{I}_4 & \mathbf{I}_4 \\ \mathbf{I}_4 & -\mathbf{I}_4 \end{bmatrix} \begin{bmatrix} \frac{1}{2}\left(\mathbf{B}_4^{(8)} + \mathbf{B}_4^{(9)}\right) & \mathbf{0} \\ \mathbf{0} & \frac{1}{2}\left(\mathbf{B}_4^{(8)} - \mathbf{B}_4^{(9)}\right) \end{bmatrix} \begin{bmatrix} \mathbf{I}_4 & \mathbf{I}_4 \\ \mathbf{I}_4 & -\mathbf{I}_4 \end{bmatrix},$$

$$\hat{\mathbf{B}}_8^{(2-)} = \begin{bmatrix} \mathbf{I}_4 & \mathbf{I}_4 \\ \mathbf{I}_4 & -\mathbf{I}_4 \end{bmatrix} \begin{bmatrix} \frac{1}{2}\left(\mathbf{B}_4^{(10)} + \mathbf{B}_4^{(11)}\right) & \mathbf{0} \\ \mathbf{0} & \frac{1}{2}\left(\mathbf{B}_4^{(10)} - \mathbf{B}_4^{(11)}\right) \end{bmatrix} \begin{bmatrix} \mathbf{I}_4 & \mathbf{I}_4 \\ \mathbf{I}_4 & -\mathbf{I}_4 \end{bmatrix},$$

where

$$\mathbf{B}_4^{(0)} + \mathbf{B}_4^{(1)} = \begin{bmatrix} d_{20}+d_{21} & d_{16}+d_{17} & d_{19}-d_{18} & d_{22}+d_{23} \\ d_4+d_5 & d_0+d_1 & d_3-d_2 & d_6+d_7 \\ d_{12}+d_{13} & d_8-d_9 & d_{10}+d_{11} & d_{14}-d_{15} \\ d_{29}-d_{28} & d_{24}+d_{25} & d_{26}+d_{27} & d_{31}-d_{30} \end{bmatrix},$$

$$\mathbf{B}_4^{(0)} - \mathbf{B}_4^{(1)} = \begin{bmatrix} d_{20}-d_{21} & d_{16}-d_{17} & d_{18}+d_{19} & d_{23}-d_{22} \\ d_4-d_5 & d_0-d_1 & d_2+d_3 & d_7-d_6 \\ d_{13}-d_{12} & d_8-d_9 & d_{10}-d_{11} & d_{14}+d_{15} \\ -d_{28}-d_{29} & d_{24}-d_{25} & d_{27}-d_{26} & d_{30}+d_{31} \end{bmatrix},$$

$$\mathbf{B}_4^{(2)} + \mathbf{B}_4^{(3)} = \begin{bmatrix} d_{30}+d_{31} & d_{27}-d_{26} & d_{24}-d_{25} & d_{28}-d_{29} \\ d_{14}+d_{15} & d_{10}-d_{11} & d_8-d_9 & d_{13}-d_{12} \\ d_6-d_7 & d_2-d_3 & d_1-d_0 & d_5-d_4 \\ d_{23}-d_{22} & d_{18}+d_{19} & d_{16}-d_{17} & d_{20}-d_{21} \end{bmatrix},$$

$$\mathbf{B}_4^{(2)} - \mathbf{B}_4^{(3)} = \begin{bmatrix} d_{30}-d_{31} & d_{26}-d_{27} & d_{24}-d_{25} & d_{28}-d_{29} \\ d_{14}-d_{15} & d_{10}+d_{11} & d_8-d_9 & d_{12}+d_{13} \\ -d_6-d_7 & d_2-d_3 & -d_0-d_1 & -d_4-d_5 \\ -d_{22}-d_{23} & d_{18}-d_{19} & d_{16}-d_{17} & d_{20}-d_{21} \end{bmatrix},$$

$$\mathbf{B}_4^{(4)} + \mathbf{B}_4^{(5)} = \begin{bmatrix} d_{10}-d_{11} & d_{14}+d_{15} & d_{13}-d_{12} & d_8-d_9 \\ d_{27}-d_{26} & d_{30}+d_{31} & d_{28}-d_{29} & d_{24}-d_{25} \\ d_{18}+d_{19} & d_{23}-d_{22} & d_{20}-d_{21} & d_{16}-d_{17} \\ -d_2-d_3 & d_6-d_7 & d_5-d_4 & d_1-d_0 \end{bmatrix},$$

$$\mathbf{B}_4^{(4)} - \mathbf{B}_4^{(5)} = \begin{bmatrix} d_{10}+d_{11} & d_{14}-d_{15} & d_{12}+d_{13} & d_8-d_9 \\ -d_{26}-d_{27} & d_{30}-d_{31} & d_{28}-d_{29} & d_{24}-d_{25} \\ d_{19}-d_{18} & d_{22}+d_{23} & d_{20}+d_{21} & d_{16}+d_{17} \\ d_3-d_2 & d_6+d_7 & d_4+d_5 & d_0+d_1 \end{bmatrix},$$

$$\mathbf{B}_4^{(6)} + \mathbf{B}_4^{(7)} = \begin{bmatrix} d_0+d_1 & d_4+d_5 & d_6+d_7 & d_3-d_2 \\ d_{16}+d_{17} & d_{20}+d_{21} & d_{22}+d_{23} & d_{19}-d_{18} \\ d_{24}+d_{25} & d_{29}-d_{28} & d_{31}-d_{30} & d_{26}+d_{27} \\ d_8-d_9 & d_{12}+d_{13} & d_{14}-d_{15} & d_{10}+d_{11} \end{bmatrix},$$

$$\mathbf{B}_4^{(6)} - \mathbf{B}_4^{(7)} = \begin{bmatrix} d_0-d_1 & d_4-d_5 & d_7-d_6 & d_2+d_3 \\ d_{16}-d_{17} & d_{20}-d_{21} & d_{23}-d_{22} & d_{18}+d_{19} \\ d_{25}-d_{24} & d_{28}+d_{29} & d_{30}-d_{31} & d_{26}-d_{27} \\ d_8+d_9 & d_{12}-d_{13} & d_{14}-d_{15} & d_{11}-d_{10} \end{bmatrix},$$

$$\mathbf{B}_4^{(8)} + \mathbf{B}_4^{(9)} = \begin{bmatrix} c_{42}-c_{43} & c_{46}-c_{47} & c_{44}-c_{45} & c_{41}-c_{40} \\ c_{36}+c_{37} & c_{32}-c_{33} & c_{34}-c_{35} & c_{38}+c_{39} \\ -c_{44}-c_{45} & c_{40}+c_{41} & c_{42}+c_{43} & c_{47}-c_{46} \\ c_{34}+c_{35} & c_{38}-c_{39} & c_{37}-c_{36} & c_{32}-c_{33} \end{bmatrix},$$

$$\mathbf{B}_4^{(8)} - \mathbf{B}_4^{(9)} = \begin{bmatrix} c_{42}+c_{43} & c_{47}-c_{46} & c_{44}-c_{45} & c_{40}+c_{41} \\ c_{36}-c_{37} & c_{33}-c_{32} & c_{34}-c_{35} & c_{39}-c_{38} \\ c_{44}-c_{45} & c_{41}-c_{40} & c_{42}-c_{43} & c_{46}-c_{47} \\ c_{34}-c_{35} & c_{38}+c_{39} & c_{36}+c_{37} & c_{32}-c_{33} \end{bmatrix},$$

$$\mathbf{B}_4^{(10)} + \mathbf{B}_4^{(11)} = \begin{bmatrix} -c_{58}-c_{59} & c_{63}-c_{62} & c_{60}-c_{61} & c_{56}-c_{57} \\ -c_{52}-c_{53} & c_{48}+c_{49} & c_{50}+c_{51} & c_{54}-c_{55} \\ c_{60}+c_{61} & c_{56}+c_{57} & c_{58}-c_{59} & c_{62}+c_{63} \\ c_{51}-c_{50} & c_{54}-c_{55} & c_{53}-c_{52} & c_{48}-c_{49} \end{bmatrix},$$

$$\mathbf{B}_4^{(10)} - \mathbf{B}_4^{(11)} = \begin{bmatrix} c_{59}-c_{58} & c_{62}-c_{63} & c_{60}-c_{61} & c_{56}-c_{57} \\ c_{53}-c_{52} & c_{48}-c_{49} & c_{51}-c_{50} & c_{54}-c_{55} \\ c_{61}-c_{60} & c_{57}-c_{56} & c_{58}+c_{59} & c_{62}-c_{63} \\ -c_{50}-c_{51} & c_{55}-c_{54} & c_{52}+c_{53} & c_{48}-c_{49} \end{bmatrix}.$$

Combining the calculations for of the all above matrices in a single procedure we finally obtain:

$$\mathbf{Y}_{32\times 1} = \mathbf{M}_{32}^{(r)}\mathbf{T}_{32\times 48}\mathbf{S}_{48}^{(r)}\mathbf{W}_{48}^{(1)}\mathbf{P}_{48}^{(r)}\mathbf{W}_{48}^{(2)}\hat{\mathbf{B}}_{48}\mathbf{W}_{48}^{(2)}\mathbf{P}_{48}^{(c)}\mathbf{W}_{48}^{(1)}\mathbf{S}_{48}^{(c)}\mathbf{T}_{48\times 32}\mathbf{M}_{32}^{(c)}\mathbf{X}_{32\times 1}, \quad (17)$$

where:

$$\hat{\mathbf{B}}_{48} = \text{quasidiag}\begin{pmatrix} \frac{1}{4}\left(\mathbf{B}_4^{(0)} + \mathbf{B}_4^{(1)}\right) \\ \frac{1}{4}\left(\mathbf{B}_4^{(0)} - \mathbf{B}_4^{(1)}\right) \\ \frac{1}{4}\left(\mathbf{B}_4^{(2)} + \mathbf{B}_4^{(3)}\right) \\ \frac{1}{4}\left(\mathbf{B}_4^{(2)} - \mathbf{B}_4^{(3)}\right) \\ \frac{1}{4}\left(\mathbf{B}_4^{(4)} + \mathbf{B}_4^{(5)}\right) \\ \frac{1}{4}\left(\mathbf{B}_4^{(4)} - \mathbf{B}_4^{(5)}\right) \\ \frac{1}{4}\left(\mathbf{B}_4^{(6)} + \mathbf{B}_4^{(7)}\right) \\ \frac{1}{4}\left(\mathbf{B}_4^{(6)} - \mathbf{B}_4^{(7)}\right) \\ \frac{1}{4}\left(\mathbf{B}_4^{(8)} + \mathbf{B}_4^{(9)}\right) \\ \frac{1}{4}\left(\mathbf{B}_4^{(8)} - \mathbf{B}_4^{(9)}\right) \\ \frac{1}{4}\left(\mathbf{B}_4^{(10)} + \mathbf{B}_4^{(11)}\right) \\ \frac{1}{4}\left(\mathbf{B}_4^{(10)} - \mathbf{B}_4^{(11)}\right) \end{pmatrix},$$

$$\mathbf{P}_{48}^{(r)} = \mathbf{I}_6 \otimes \mathbf{P}_8^{(r)},$$

$$\mathbf{P}_{48}^{(c)} = \mathbf{I}_6 \otimes \mathbf{P}_8^{(c)},$$

$$\mathbf{W}_{48}^{(2)} = \mathbf{I}_6 \otimes \mathbf{H}_2 \otimes \mathbf{I}_4.$$

Figure 1 shows a data flow diagram describing the new algorithm for the computation of the product of Kaluza numbers (17). In this paper, the data flow diagram is oriented from left to right. Straight lines in the figure denote the operations of data transfer. Points, where lines converge, denote summation. The dotted lines indicate the subtraction operation. We use the regular lines without arrows on purpose, so as not to clutter the picture. The rectangles indicate the matrix-vector multiplications with matrices inscribed inside a rectangle.

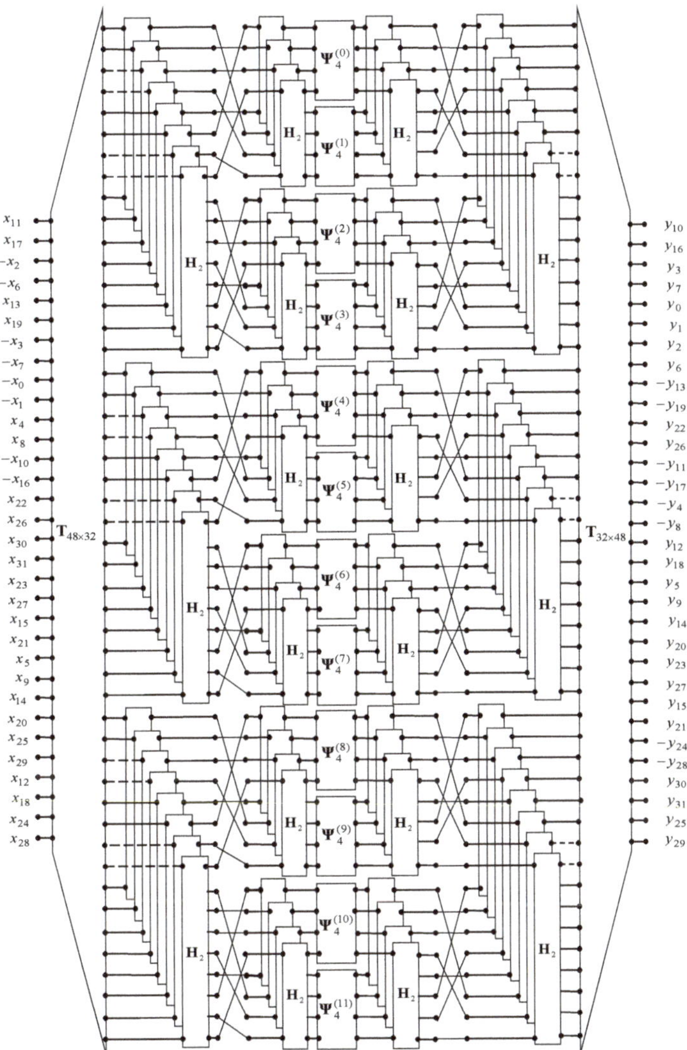

Figure 1. A data flow diagram for the proposed algorithm.

4. Evaluation of Computational Complexity

We will now calculate how many multiplications and additions of real numbers are required for the implementation of the new algorithm and will compare this with the number of operations required both for direct computation of matrix-vector products in Equation (1) and for implementing our previous algorithm [25]. The number of real multiplications required using the new algorithm is 192. Thus, using the proposed algorithm, the number of real multiplications needed to calculate the Kaluza number product is significantly reduced. The number of real additions required using our algorithm is 384. We observe that the direct computation of the Kaluza number product requires 608 additions more than the proposed algorithm. Thus, our proposed algorithm saves 832 multiplications and 960 additions of real numbers compared with the direct method. Thus, the total number of arithmetic operations for the proposed algorithm is approximately 71.4% less than that of the direct computation. The previously proposed algorithm [25] calculates the same result using 512 multiplications and 576 additions of real numbers. Thus, our proposed algorithm saves 62.5% of multiplications and 33.3% of additions of real numbers compared with our previous algorithm. Hence, the total number of arithmetic operations for the new proposed algorithm is approximately 47% less than that of our previous algorithm.

5. Conclusions

We presented a new effective algorithm for calculating the product of two Kaluza numbers. The use of this algorithm reduces the computational complexity of multiplications of Kaluza numbers, thus reducing implementation complexity and leading to a high-speed resource-effective architecture suitable for parallel implementation on VLSI platforms. Additionally, we note that the total number of arithmetic operations in the new algorithm is less than the total number of operations in the compared algorithms. Therefore, the proposed algorithm is better than the compared algorithms, even in terms of its software implementation on a general-purpose computer.

The proposed algorithm can be used in metacognitive neural networks using Kaluza numbers for data representation and processing. The effect in this case is achieved by using non-commutative finite groups based on the properties of the hypercomplex algebra [24]. When using the Kaluza number, in this case, the rule for generating the elements of the group will be set, as well as the rule for performing the group operation of multiplication. Such a system can contain two components: a neural network based on Kaluza numbers, which represents a cognitive component, and a metacognitive component, which serves to self-regulate the learning algorithm. At each stage, the metacognitive component will decide how and when the learning takes place. The algorithm removes unnecessary samples and keeps only those that are used. This decision will be determined by the magnitude and 31 phases of the Kaluza number. However, these matters are beyond the scope of this article and require more detailed research.

Author Contributions: Conceptualization, A.C.; methodology, A.C., G.C. and J.P.P.; validation, J.P.P.; formal analysis, A.C. and J.P.P.; writing—original draft preparation, A.C. and J.P.P.; writing—review and editing, A.C. and J.P.P.; visualization, A.C. and J.P.P.; supervision, A.C., G.C. and J.P.P. All authors have read and agreed to the published version of the manuscript.

Funding: This research received no external funding.

Conflicts of Interest: The authors declare no conflict of interest.

References

1. Kantor, I.L.; Solodovnikov, A.S. *Hypercomplex Numbers: An Elementary Introduction to Algebras*; Springer: Berlin/Heidelberg, Germany, 1989.
2. Alfsmann, D. On families of 2^N-dimensional hypercomplex algebras suitable for digital signal processing. In Proceedings of the 2006 14th European Signal Processing Conference, Florence, Italy, 4–8 September 2006; pp. 1–4.
3. Alfsmann, D.; Göckler, H.G.; Sangwine, S.J.; Ell, T.A. Hypercomplex algebras in digital signal processing: Benefits and drawbacks. In Proceedings of the 2007 15th European Signal Processing Conference, Poznan, Poland, 3–7 September 2007; pp. 1322–1326.
4. Bayro-Corrochano, E. Multi-resolution image analysis using the quaternion wavelet transform. *Numer. Algorithms* **2005**, *39*, 35–55. [CrossRef]
5. Belfiore, J.C.; Rekaya, G. Quaternionic lattices for space-time coding. In Proceedings of the 2003 IEEE Information Theory Workshop (Cat. No. 03EX674), Paris, France, 31 March–4 April 2003; pp. 267–270.
6. Bulow, T.; Sommer, G. Hypercomplex signals-a novel extension of the analytic signal to the multidimensional case. *IEEE Trans. Signal Process.* **2001**, *49*, 2844–2852. [CrossRef]
7. Calderbank, R.; Das, S.; Al-Dhahir, N.; Diggavi, S. Construction and analysis of a new quaternionic space-time code for 4 transmit antennas. *Commun. Inf. Syst.* **2005**, *5*, 97–122.
8. Ertuğ, Ö. Communication over Hypercomplex Kähler Manifolds: Capacity of Multidimensional-MIMO Channels. *Wirel. Person. Commun.* **2007**, *41*, 155–168. [CrossRef]
9. Le Bihan, N.; Sangwine, S. Hypercomplex analytic signals: Extension of the analytic signal concept to complex signals. In Proceedings of the 15th European Signal Processing Conference (EUSIPCO-2007), Poznan, Poland, 3–7 September 2007; p. A5P–H.
10. Moxey, E.C.; Sangwine, S.J.; Ell, T.A. Hypercomplex correlation techniques for vector images. *IEEE Trans. Signal Process.* **2003**, *51*, 1941–1953. [CrossRef]

11. Comminiello, D.; Lella, M.; Scardapane, S.; Uncini, A. Quaternion convolutional neural networks for detection and localization of 3d sound events. In Proceedings of the ICASSP 2019–2019 IEEE International Conference on Acoustics, Speech and Signal Processing (ICASSP), Brighton, UK, 12–17 May 2019; pp. 8533–8537.
12. de Castro, F.Z.; Valle, M.E. A broad class of discrete-time hypercomplex-valued Hopfield neural networks. *Neural Netw.* **2020**, *122*, 54–67. [CrossRef]
13. Gaudet, C.J.; Maida, A.S. Deep quaternion networks. In Proceedings of the 2018 International Joint Conference on Neural Networks (IJCNN), Rio de Janeiro, Brazil, 8–13 July 2018; pp. 1–8.
14. Isokawa, T.; Kusakabe, T.; Matsui, N.; Peper, F. Quaternion neural network and its application. In Proceedings of the International Conference on Knowledge-Based and Intelligent Information and Engineering Systems, Oxford, UK, 3–5 September 2003; Springer: Berlin/Heidelberg, Germany, 2003; pp. 318–324.
15. Liu, Y.; Zheng, Y.; Lu, J.; Cao, J.; Rutkowski, L. Constrained quaternion-variable convex optimization: A quaternion-valued recurrent neural network approach. *IEEE Trans. Neural Netw. Learn. Syst.* **2019**, *31*, 1022–1035. [CrossRef]
16. Parcollet, T.; Morchid, M.; Linarès, G. A survey of quaternion neural networks. *Artif. Intell. Rev.* **2020**, *53*, 2957–2982. [CrossRef]
17. Saoud, L.S.; Ghorbani, R.; Rahmoune, F. Cognitive quaternion valued neural network and some applications. *Neurocomputing* **2017**, *221*, 85–93. [CrossRef]
18. Saoud, L.S.; Ghorbani, R. Metacognitive octonion-valued neural networks as they relate to time series analysis. *IEEE Trans. Neural Netw. Learn. Syst.* **2019**, *31*, 539–548. [CrossRef] [PubMed]
19. Vecchi, R.; Scardapane, S.; Comminiello, D.; Uncini, A. Compressing deep-quaternion neural networks with targeted regularisation. *CAAI Trans. Intell. Technol.* **2020**, *5*, 172–176. [CrossRef]
20. Vieira, G.; Valle, M.E. A General Framework for Hypercomplex-valued Extreme Learning Machines. *arXiv* **2021**, arXiv:2101.06166.
21. Wu, J.; Xu, L.; Wu, F.; Kong, Y.; Senhadji, L.; Shu, H. Deep octonion networks. *Neurocomputing* **2020**, *397*, 179–191. [CrossRef]
22. Zhu, X.; Xu, Y.; Xu, H.; Chen, C. Quaternion convolutional neural networks. In Proceedings of the European Conference on Computer Vision (ECCV), Munich, Germany, 8–14 September 2018; pp. 631–647.
23. Bojesomo, A.; Liatsis, P.; Marzouqi, H.A. Traffic flow prediction using Deep Sedenion Networks. *arXiv* **2020**, arXiv:2012.03874.
24. Saoud, L.S.; Al-Marzouqi, H. Metacognitive sedenion-valued neural network and its learning algorithm. *IEEE Access* **2020**, *8*, 144823–144838. [CrossRef]
25. Cariow, A.; Cariowa, G.; Łentek, R. An algorithm for multipication of Kaluza numbers. *arXiv* **2015**, arXiv:1505.06425.
26. Silvestrov, V.V. Number Systems. *Soros Educ. J.* **1998**, *8*, 121–127.
27. Ţariov, A. *Algorithmic Aspects of Computing Rationalization in Digital Signal Processing. (Algorytmiczne Aspekty Racjonalizacji Obliczeń w Cyfrowym Przetwarzaniu Sygnałów)*; West Pomeranian University Press: Szczecin, Poland, 2012. (In Polish)
28. Ţariov, A. Strategies for the synthesis of fast algorithms for the computation of the matrix-vector products. *J. Signal Process. Theory Appl.* **2014**, *3*, 1–19.

Article

System for Neural Network Determination of Atrial Fibrillation on ECG Signals with Wavelet-Based Preprocessing

Pavel Lyakhov [1,2,*], Mariya Kiladze [1] and Ulyana Lyakhova [1]

1 Department of Mathematical Modeling, North Caucasus Federal University, Pushkin Str. 1, 355017 Stavropol, Russia; merchali@mail.ru (M.K.); uljahovs@mail.ru (U.L.)
2 North-Caucasus Center for Mathematical Research, North-Caucasus Federal University, Pushkin Str. 1, 355017 Stavropol, Russia
* Correspondence: ljahov@mail.ru; Tel.: +79-620-287-214

Featured Application: The use by medical of a neural network classification system for ECG signals with preprocessing steps to recognize atrial fibrillation.

Abstract: Today, cardiovascular disease is the leading cause of death in developed countries. The most common arrhythmia is atrial fibrillation, which increases the risk of ischemic stroke. An electrocardiogram is one of the best methods for diagnosing cardiac arrhythmias. Often, the signals of the electrocardiogram are distorted by noises of varying nature. In this paper, we propose a neural network classification system for electrocardiogram signals based on the Long Short-Term Memory neural network architecture with a preprocessing stage. Signal preprocessing was carried out using a symlet wavelet filter with further application of the instantaneous frequency and spectral entropy functions. For the experimental part of the article, electrocardiogram signals were selected from the open database PhysioNet Computing in Cardiology Challenge 2017 (CinC Challenge). The simulation was carried out using the MatLab 2020b software package for solving technical calculations. The best simulation result was obtained using a symlet with five coefficients and made it possible to achieve an accuracy of 87.5% in recognizing electrocardiogram signals.

Keywords: digital filter; electrocardiogram; instantaneous frequency; symlet wavelet; spectral entropy; signal denoising; LSTM

Citation: Lyakhov, P.; Kiladze, M.; Lyakhova, U. System for Neural Network Determination of Atrial Fibrillation on ECG Signals with Wavelet-Based Preprocessing. *Appl. Sci.* **2021**, *11*, 7213. https://doi.org/10.3390/app11167213

Academic Editor: Fabio La Foresta

Received: 11 June 2021
Accepted: 2 August 2021
Published: 5 August 2021

1. Introduction

The number of people who suffer from cardiac diseases is increasing every day. This disease is the leading cause of death in developed countries [1–3]. Electrocardiography is a method of recording and studying the electric fields that are generated during the work of the heart. An ECG is the result of electrocardiography [4,5] and is a graphical record of the electrical activity of the heart produced by depolarization and repolarization of the atria and ventricles. The electrocardiogram (ECG) is a non-invasive technique used to detect cardiovascular disease. The ECG is described by the waveforms of the P, QRS, and T waves, which are associated with each heart-rate function. The P wave displays the process of depolarization of the atrial myocardium; the QRS complex displays depolarization of the ventricles; the ST segment, and the T wave displays the processes of repolarization of the ventricular myocardium [6]. Figure 1 shows an example of an electrocardiogram waveform with P, Q, R, S, and T characteristics, as well as standard electrocardiogram intervals are PQ intervals, ST intervals, and QRS complex.

Atrial fibrillation is a major risk factor for ischemic stroke [7]. The main criteria for the presence of atrial fibrillation are the absence of P waves, the presence of atrial fibrillation waves, different R-R intervals, the heart-rate (HR) is constant or accelerated, and the QRS complex is less than 0.12 s [8]. Since the P wave is not detected during atrial fibrillation, the interval between QRS complexes increases and there is no possibility to calculate the

PQ and QT intervals. Calculating HR or QRS time from digital ECG signals is problematic. Therefore, it is necessary to pay attention to the absence of the P-peak and the presence of different intervals between R-R peaks when determining atrial fibrillation.

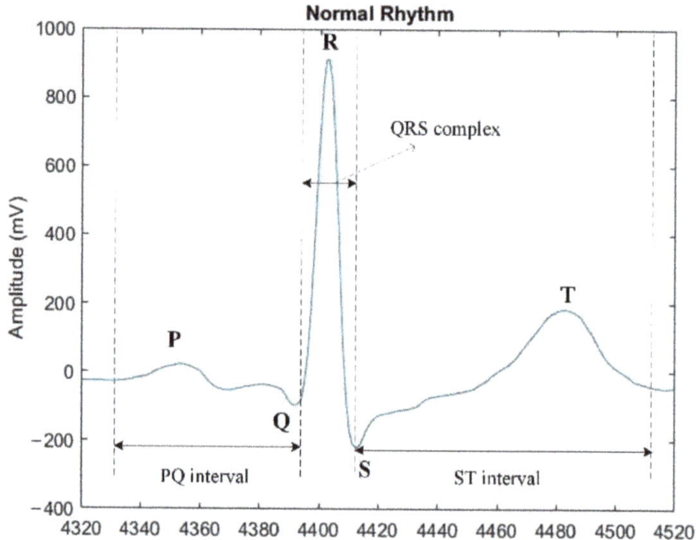

Figure 1. Example of an electrocardiogram signal.

In some modern electrocardiographs, various signal filters are used, which allow obtainment of a higher quality of the electrocardiogram, while introducing some distortions in the form of the received signal. Low-pass filters in the range of 0.5–1 Hz can reduce the effect of the floating contour while introducing distortions in the shape of the ST segment [4]. A low-frequency anti-tremor filter in the 35 Hz range suppresses artifacts associated with muscle activity. A notch filter in the range of 50–60 Hz neutralizes line pickups [5].

Today, learning algorithms are becoming more accurate, but recognition systems created based on artificial intelligence in medicine, and in particular, in cardiology, are not able to achieve a 100% accurate result [9]. For this reason, it is relevant to search for ways to increase this indicator. One of the possible ways to increase this indicator is the preliminary processing of ECG signals. Finding peaks in signals is an important step in many automatic ECG processing systems. In this work, a neural network classification system for ECG signals is proposed for determining atrial fibrillation with a preprocessing stage. At the stage of preprocessing, signals are selected by the number of heartbeat counts, as well as wavelet analysis to clean up noise and isolate the R-peak and spectral analysis to isolate the P-peak. Automatic detection of atrial fibrillation from ECG signals will allow doctors to determine if a patient needs cardiac care.

2. Related Research

Today, medicine is considered one of the promising areas for the introduction of artificial intelligence, because the analysis of medical signals is the most common research method in this area. An example is an algorithm for identifying patients with atrial fibrillation in sinus rhythm based on convolutional neural networks (CNN) in [10]. The work [11] presents a hybrid approach with the use of Empirical Mode Decomposition and CNN to classify ECG signals. In [12], the author has implemented the extraction of ECG signal features based on wavelet transform with further classification using Long Short-Term Memory (LSTM). The work [13] presents a method for detecting atrial fibrillation using LSTM. The method achieved an accuracy of 98.51%. The essence of the approach proposed in the work is the separation of ECG signals with a sliding window and the loading of the obtained blocks into the decision-making system. ECG signals were taken from the MIT-BIH database Atrial Fibrillation Database. Dataset was 10 h long and each signal contained 100 heartbeats (R-R peak). The authors point out the slow learning speed and high requirements for computing resources. For the simulation, the authors used the Quadro M5000 GPU (Nvidia, Santa Clara, CA, USA).

This work is of scientific interest, however, due to the different methodology and resources used, it cannot be compared with the proposed system for neural network determination of atrial fibrillation on ECG signals with wavelet-based preprocessing.

Finding peaks in signals is an important step in many signal processing applications. Automatic peak detection using neural network classification systems is difficult due to the physiological variability of P waves and QRS complexes, as well as the presence of various types of noise, including muscle noise, artifacts due to electrode movement, power-line noise, and baseline deviations.

Software QRS complex recognition is an integral part of modern computerized ECG monitoring systems. An algorithm for their recognition is presented in [14] and is based on optimized filtering and simple threshold setting since optimized filtering is considered a factor in achieving good timing accuracy. The most widely known method for detecting a single R-peak is the Pan-Tomkins method, which uses three types of processing steps: linear digital filtering, nonlinear transformation, and decision rule algorithms [15]. The work [16] presents an algorithm for detecting QRS using multi-stage morphological filtering to suppress impulse noise. In work [17], to determine the QRS complex, it is proposed to create an estimated QRS signal using the parameters extracted from the original ECG signal.

For automatic disease diagnosis systems based on ECG signals, accurate determination of the P wave is critical. In work [18], one of the first methods of processing ECG signals to isolate the P-peak in its flow using wavelet transform and subsequent training of the neural network is described. In [3], a system for determining the P and T waves based on the wavelet transform is presented.

The main methods of processing ECG signals for noise reduction are digital filtering, adaptive filtering, wavelet filtering. The paper [19] describes a method for processing ECG signals using an adaptive wavelet transform based on the Poincare section and the Shannon method. There are also methods for processing signals online in real-time for use in pacemakers. The paper [20] describes a processing method using biorthogonal wavelet transform based on a linear phase structure for noise removal, feature extraction, and compression of the ECG signal. There are also methods for computerized detection of fibrillation. Work [21] describes a method for detecting fibrillation using an eight-layer neural convolutional network, which requires only basic data normalization without preliminary processing and extraction of features from raw ECG samples. Work [7] describes a method for constructing classifiers based on several sets of functions (a set of Andreotti, Zabikhi functions, an aggregated set of functions, and a set of Dutt functions) and using a random forest classification method.

3. Materials and Methods

3.1. Neural Network System for Atrial Fibrillation Recognition by ECG Signal

Any digital signal is distorted by noises of varying nature. Noise on the ECG signals makes it difficult to analyze the data, both for the specialist and for systems based on artificial intelligence. Signal preprocessing allows the ECG to be prepared for further classification. This paper proposes a system for determining atrial fibrillation by ECG signals, which includes four stages presented in Figure 2. The first stage consists of the preprocessing of signals to isolate signals with the same count of heartbeats. At the second stage, noise is removed from the ECG signals using a discrete wavelet transform. At the third stage, the P-peak is isolated using spectral analysis. The fourth step is to classify signals using the LSTM network.

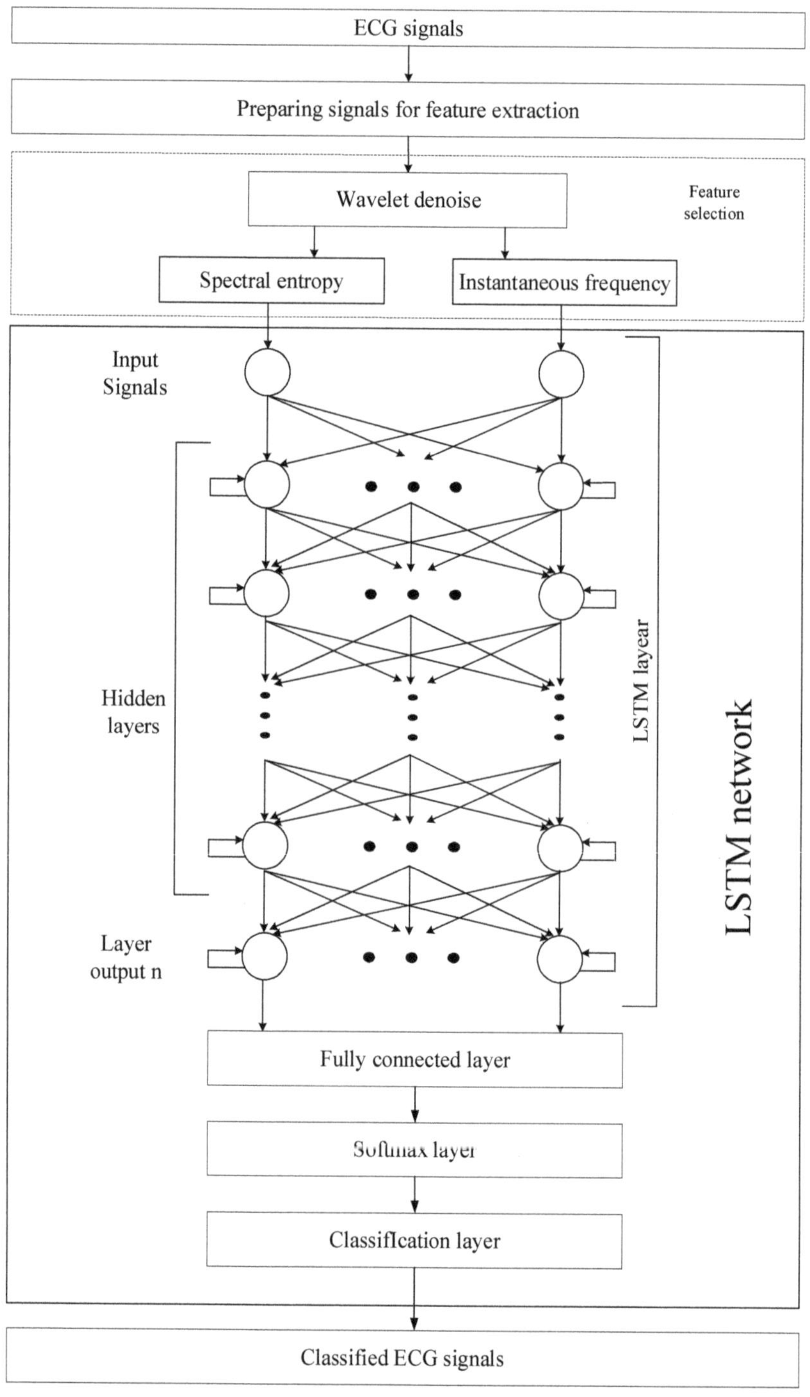

Figure 2. Neural network classification system with the pre-processing stage for ECG signals.

3.2. Method for Pre-Processing of ECG Signals

As part of the preprocessing of the ECG signal, its length is checked. Each ECG signal has a specific number of heartbeats and several samples. The length of ECG signals is measured in the number of samples that make up the signal. The ECG signal database can contain signals with a different number of samples. For the correct operation of the neural network classification system, the number of samples must be the same for all signals. To select ECG signals with one length, the following steps are necessary. At the first stage, the ECG signals must be divided into groups with the same number of samples. The second step is to select a group consisting of the largest number of signals of the same length. The third step removes signals consisting of fewer and more samples than in the selected group.

3.3. Removing Noise from ECG Signals Using a Discrete Wavelet Transform

Any digital signal is distorted by noises of varying nature. To isolate signal features, it is necessary to clean noise from them. An ECG signal that is distorted by noise can be written as:

$$W(t) = S(t) + N(t), \tag{1}$$

where $W(t)$ is the ECG signal, $S(t)$ is the ECG signal without noise distortion, $N(t)$ is the noise on the ECG signal.

Wavelet transform is a common way to remove noise from a signal [22]. To clean the noise from the ECG signals, a discrete wavelet transform (DWT) of the symlet family was used, which is a Daubechies wavelet with the least asymmetry and a compact carrier. The detail factor for each case is set empirically.

There are three stages of using wavelet transform to clean the noise from the ECG signal. The first step is to obtain noisy wavelet coefficients using the DWT of a noisy signal. The second stage is the choice of thresholding. The third stage is an inverse wavelet transform to obtain a purified signal [23]. The DWT of the ECG signal is:

$$DWT(a,b) = \frac{1}{\sqrt{2}} \sum_{j=0}^{N} W_j \int_{j}^{j+1} \psi\left(\frac{t-b}{a}\right) dt, \tag{2}$$

where N is the number of samples on the ECG signal, W is the ECG signal distorted by noise, ψ is a symlet, a and b variables can take on the values $a = 1 \ldots N, b = 1 \ldots N-1$.

To obtain DWT, a low-pass analysis filter with an g impulse response and a high-pass analysis filter with an h impulse response are used. As a result of filtering, approximating and detailing coefficients are obtained [24].

$$y_{low}(t) = \sum_{k=-\infty}^{\infty} W(k) g(2t-k), \tag{3}$$

$$y_{high}(t) = \sum_{k=-\infty}^{\infty} W(k) h(2t-k). \tag{4}$$

With each DWT application, the number of samples on the ECG signal is halved by Table 1 [24,25]. Each subsequent decomposition is carried out in terms of the low-frequency component by Figure 3.

Table 1. Length of ECG signals in DWT, where N is the initial number of samples on the ECG signal.

Number of DWT Levels	Number of Samples on the ECG Signal
1	$\frac{N}{2^1}$
2	$\frac{N}{2^2}$
...	...
n	$\frac{N}{2^n}$

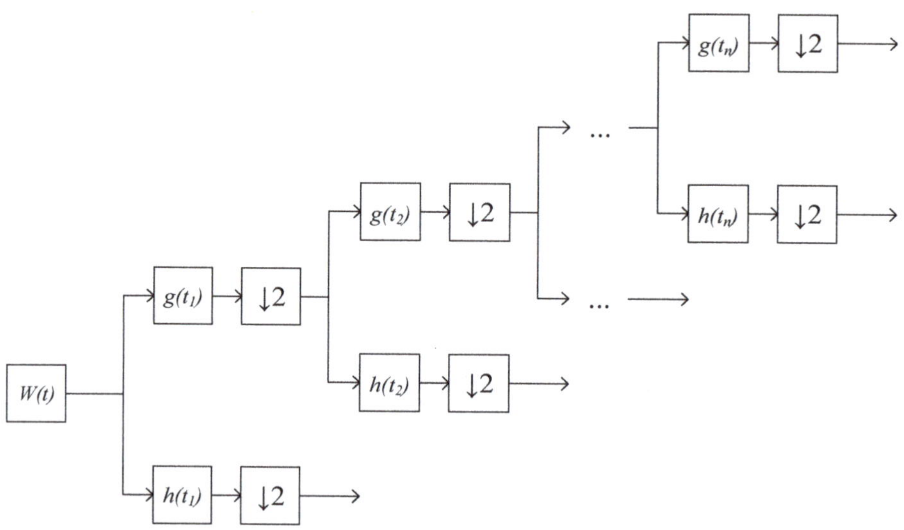

Figure 3. DWT ECG signal, where $W(t)$ is an ECG signal distorted by noise, $2\downarrow$ is a decimation, g is a low-pass analysis filter, h is a high-pass analysis filter.

The next step in clearing noise from the ECG signal is to select a threshold function and thresholding. The thresholding is a value that determines whether there is noise in the signal. If the value of the wavelet coefficient at a certain moment is greater than the value of the threshold, then it is considered the value of the signal, if less, then the noise [24]. To determine the threshold limit of the ECG signal, the minimax threshold was used [26]:

$$T \leq \sqrt{2\ln N},\ T^2 = 2\ln(N+1) - 4\ln(\ln(N+1)) - \ln 2\pi \tag{5}$$

The wavelet coefficients are transformed by thresholding, using the threshold functions. The functions of hard- and soft-threshold determination are most often used [24]. To determine the noise on the ECG signal, the soft-threshold function was used [26]:

$$0 \leq \max\left(1 - \frac{T}{|x|}, 0\right) \leq 1, \tag{6}$$

where x is the value of the wavelet coefficient.

The reverse DWT for obtaining a cleaned ECG signal is as follows [24]:

$$S(t) = \sum_{k=-\infty}^{\infty} \widetilde{g}_k y_{low_k}(t) + \sum_{m=-\infty}^{\infty} \sum_{k=-\infty}^{\infty} \widetilde{h}_{m_k} y_{high_k}(t), \tag{7}$$

where $S(t)$ is the ECG signal without noise distortion, $\widetilde{g}_k$ and $\widetilde{h}_{m_k}$ are approximating and detailing coefficients after processing by the threshold function. The reverse DWT of the ECG signal is performed according to Figure 4.

3.4. Isolation of the P-Peak Feature Using Spectral Analysis

The instantaneous frequency and spectral entropy functions were selected to isolate the P-peak. To calculate it, it is necessary to calculate the amplitude spectrum of the process using the Fourier transform, then normalize the amplitude spectrum so that the sum of its readings becomes equal to 1 and calculate the entropy using Shannon's formula. Changes in spectral entropy over time are associated with changes in the waveform, which allows its use to distinguish features on the ECG signal. Since the ECG signal consists of a finite number of samples, Shannon's formula for calculating the spectral entropy of the ECG signal is:

$$W(t) = -\sum_{i=1}^{N} n_i \log n_i, \quad (8)$$

where S is the amount of information, N is the number of possible events, n_i is the value of the i-th samples on the ECG signal. However, it is more correct to use the Fourier transform when working with stationary signals. Therefore, for more accurate identification of the P-peak on the ECG signal, several signal-processing methods are required. With a deviation in the work of the heart, changes in the frequency of the ECG signals occur. To determine such changes, an instantaneous frequency is used, since this method allows the researcher to take into account the nature of the process, which changes over time [27].

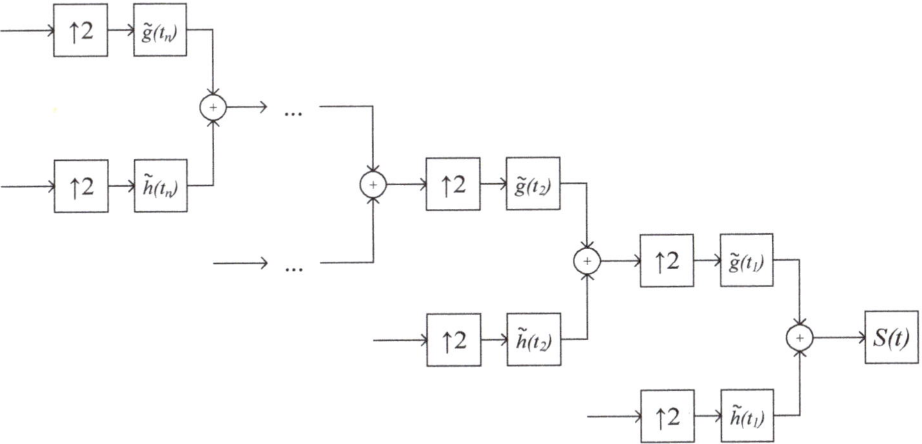

Figure 4. Reverse DWT ECG signal, where $S(t)$ is a noise-free ECG signal, $2\uparrow$ is interpolation, $\tilde{g}$ is a low-pass synthesis filter, $\tilde{h}$ is a high-pass synthesis filter.

As the ECG signal is non-stationary, in order to calculate the instantaneous frequency, the researcher can refer to the works of Carson and Fry [28] and Van de Pol [29]:

$$f_i(t) = \frac{1}{2\pi} \frac{dW(t)}{dt}, \quad (9)$$

where $W(t)$ is an ECG signal. Formula (9) describes the rate of change in the phase of the ECG signal, i.e., the instantaneous frequency shows how often the peaks appear and disappear.

3.5. LSTM Processing of ECG Data

LSTM networks have been specifically designed to find patterns over time [30]. Since ECG signals are sequences of peaks, the ability to memorize characteristic fragments of time series is critical when using LSTM in this area. Long-term and short-term memory is the main reason that the LSTM network is used as the basic structure of ECG signal recognition systems. In the present study, the deep LSTM network was used to classify ECG signals.

Passing a signal through a standard LSTM network structure involves four stages. The first stage is the passage of the signal through the sigmoidal layer, which is designed to determine the desired information. The second stage consists of passing through the sigmoidal layer, which determines whether this or that information is relevant and transmits a signal to the hyperbolic layer, which defines a new vector of candidates. The third stage is to save a new vector of candidates. The fourth stage consists of passing through the sigmoidal layer, which is designed to determine the required information, and the hyperbolic tangent to display information in the range [−1; 1]. Figure 5 shows a typical LSTM structure.

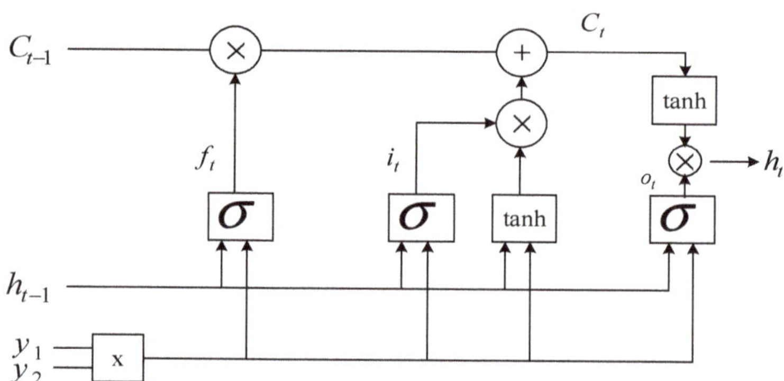

Figure 5. Standard LSTM structure.

To classify ECG signals, it is proposed to use an LSTM network with two input layers y_1 and y_2, which are combined into a two-dimensional vector x. All calculations are performed according to the standard LSTM structure, which is shown in Figure 3, where x is the input two-dimensional vector, h_{t-1} is the output vector from the previous LSTM ($h_0 = 0$) block, h_t is the output vector of the LSTM block, C_{t-1} is the state vector from the previous LSTM ($C_0 = 0$) block, C_t is the state vector of the LSTM block, σ is the sigmoid activation function, *tanh* is the hyperbolic tangent activation function, $\times$ is the multiplication operator, $+$ is the addition operator.

The forget gate vector f_t is the result of a computational step through the sigmoidal layer, which is intended to determine the desired information. The resulting vector determines what information needs to be "memorized" and is calculated by the formula:

$$f_t = \sigma\left(M_f[h_{t-1}, x] + b_f\right), \tag{10}$$

where M_f is a matrix of parameters, b_f is a vector of parameters. Update gate vector i_t checks the relevance of the information using the sigmoidal activation function:

$$i_t = \sigma(M_i[h_{t-1}, x] + b_i), \tag{11}$$

where M_i is a matrix of parameters, b_i is a vector of parameters.

The next step is to define a new state vector.

$$C_t = f_t \times C_{t-1} + i_t \times tanh(M_C[h_{t-1}, x] + b_C), \tag{12}$$

where M_C is a matrix of parameters, b_c is a vector of parameters, $\times$ is the multiplication operator.

The output gate vector O_t that is a candidate for leaving the LSTM network is calculated by the formula:

$$O_t = \sigma(M_O[h_{t-1}, x] + b_O), \tag{13}$$

where M_O is a matrix of parameters, b_O is a vector of parameters. The definition of the output vector h_t is made according to the formula:

$$h_t = O_t \times \sigma(C_t). \tag{14}$$

The rest of the LSTM network layers are standard and are used in neural networks to classify signals.

4. Results

For modeling, ECG signals were selected from the international open database PhysioNet Computing in Cardiology Challenge 2017 (CinC Challenge) [31]. This database contains over 10,000 ECG records; it is freely available from AliveCor and is a random sample of patient records of no more than one minute in duration. The Physionet Computing in Cardiology Challenge 2017 database consists of 8528 ECG signals for training and 3658 ECG signals for validation. The base consists of four types of single-channel signals: 5152 normal signals (N), 771 signals with cardiac fibrillation (A), 46 noisy signals (~) and 2557 other signals (O). The simulations were performed using two categories of signals, namely signals without heart defects (N) and signals with atrial fibrillation (A). These signal categories were selected to study the signs of atrial fibrillation on ECG signals for more correct LSTM learning. A total of 1000 signals were selected from the database CinC Challenge for the first modeling. For the second experimental simulation, 5925 ECG signals were selected, namely 5152 normal signals (N) and 771 signals with cardiac fibrillation (A). Examples of selected ECG signals from the database CinC Challenge are shown in Figure 6.

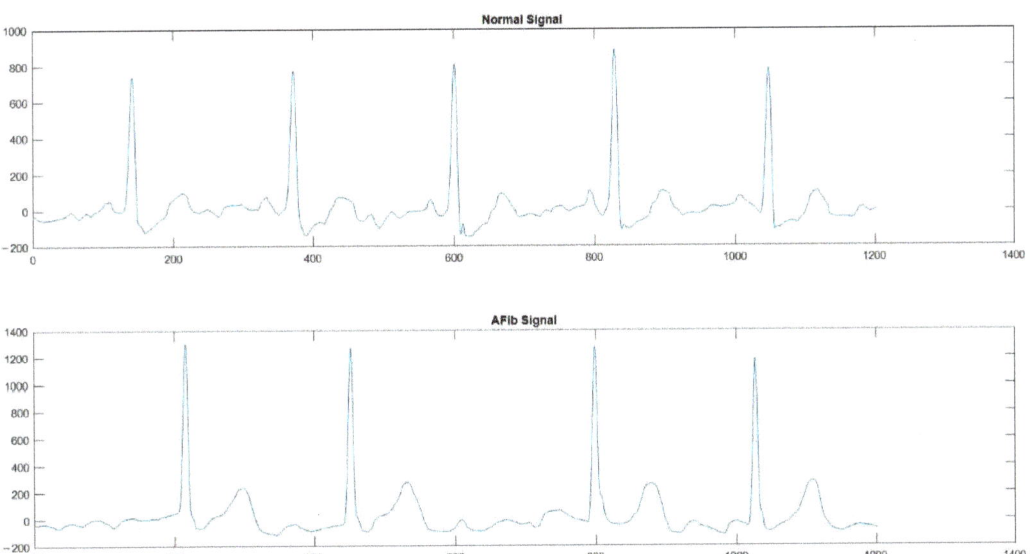

Figure 6. An example of ECG signals without heart defects (Normal Signal) and signals with atrial fibrillation (AFib signal) from the CinC Challenge database [31].

The simulation was carried out using the MatLab 2020b software package for solving technical calculations. The calculations were performed on a PC with a processor Intel(R) Core (TM) i5-10210U CPU @ 1.60 GHz (8 CPUs), 2.1 GHz.

Database PhysioNet Computing in Cardiology Challenge 2017 (CinC Challenge) consists of ECG signals with different numbers of samples. Figure 7 shows that there are significantly more signals from 9000 samples. For correct training of the neural network, the input ECG signals must contain the same number of samples. Therefore, at the stage of preliminary data processing for the first training, 976 ECG signals with several samples to 9000 were selected. For the second simulation, 5754 ECG signals with 9000 samples were taken at the preprocessing stage.

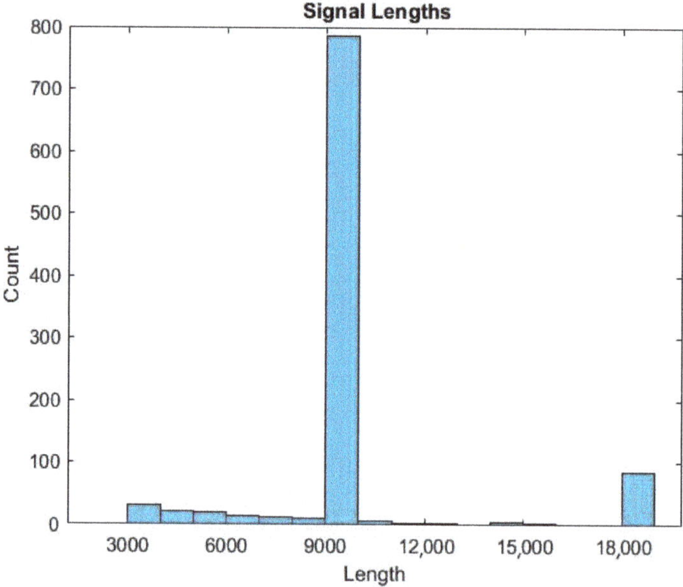

Figure 7. Histogram of the ratio of the number of samples to the number of signals.

DWT allows splitting the signal into high and low frequencies. Analysis of high frequencies of the ECG signal can determine the presence of peaks. Analysis of low frequencies of the ECG signal allows determining the presence of noise of varying nature [22]. The symlet is an orthogonal wavelet and can be used to reconstruct the signal [23] or to find R-peaks on ECG signals [24,32]. To remove noise from ECG signals and isolate R-peaks on them, the symlet wavelet filter was chosen. The symlet is similar in shape to the QRS complex on the ECG signal. This means that the decomposition coefficients of the ECG signal using the symlet-based DWT have a high correlation with the location of the P-peaks. Their number can be determined based on the analysis of the coefficients of the wavelet decomposition of the ECG signal. Figure 8 shows a graphical display of the symlet and QRS complex. Figure 9 shows examples of the five-level decomposition of an ECG signal based on a symlet. The spikes in coefficients D 2, D 3 and D 4 correspond to heartbeats, from which it is possible to determine the location of the R-peaks of the cardiogram.

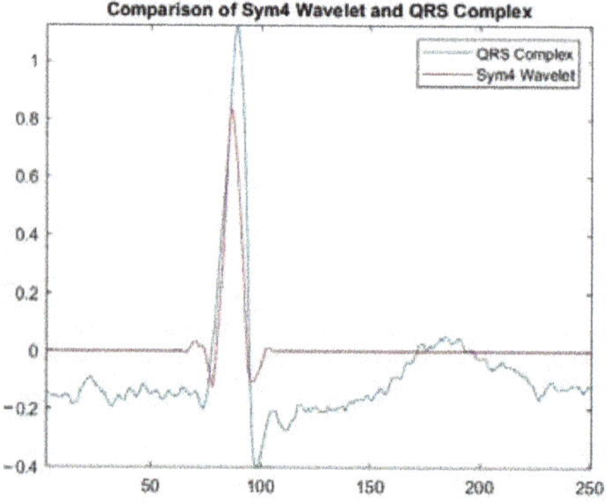

Figure 8. Symlet repeating the shape of the R-peak on the ECG signal.

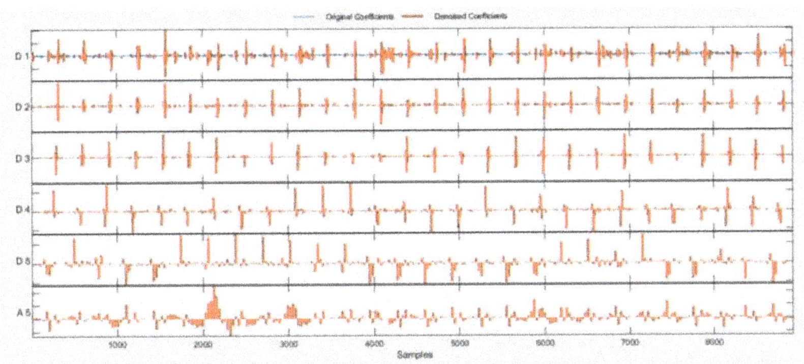

Figure 9. An example of a five-level decomposition of an ECG signal based on a symlet.

The results from Table 2 were obtained by simulating the proposed system using various symlet. Each value of the table is the result of training the proposed system for neural network determination of atrial fibrillation based on LSTM with signal preprocessing using different coefficients of symlet. The best learning result was obtained using a five-coefficient symlet.

Table 2. Simulation of the proposed method using a wavelet symlet with different coefficients.

Wavelet	Learning Outcome, %
symlet 2	66.1
symlet 3	71.0
symlet 4	78.0
symlet 5	**87.5**
symlet 6	82.1

For the correct selection of various signs on the ECG signals, one-dimensional functions must be used. The functions of instantaneous frequency and spectral entropy were selected to isolate the P-peak. Fourier series and integral Fourier transform are the basis of harmonic signal analysis. However, when analyzing ECG signals, these transformations do not provide the possibility of analyzing peaks, understanding the local properties of the signal and its frequency characteristics. Therefore, for these purposes, we used the characteristics of the Fourier spectrum. Instantaneous frequency calculates a spectrogram using short-term Fourier transforms versus window time. Spectral entropy estimates entropy based on a power spectrogram. The time output of the function corresponds to the center of the time windows.

The selected ECG signals from the CinC Challenge database were divided into signals for training and signals for testing in a percentage ratio of 90:10. For training, the architecture of the LSTM neural network was assembled. The network consisted of two input layers, to which preprocessed signals were applied, and one hundred hidden recurrent layers. Preprocessing of signals using a symlet and subsequent application of spectral analysis functions made it possible to reduce the length of ECG signals to 255 HC. Table 3 presents the results of modeling various methods for detecting atrial fibrillation. The matrix of inaccuracies as a result of training the proposed system of neural network determination of atrial fibrillation from ECG signals is presented in Figures 10 and 11.

The best indicator of the accuracy of ECG signal recognition was obtained using the method proposed in the work and amounted to 87.5%. This accuracy was obtained in the first simulation using 976 ECG signals. The indicator of the accuracy of ECG signal recognition during the second simulation was 87.4%. This result is identical to that obtained in the first simulation. The smallest indicator of recognition accuracy was obtained using a one-dimensional LSTM without preliminary signal processing. The results obtained indicate that the use of pre-processing of ECG signals can significantly increase the recognition accuracy of neural network classification systems.

Table 3. Simulation results of various methods for detecting atrial fibrillation.

	ECG Processing Method	Accuracy of Atrial Fibrillation Recognition on the ECG Signal, %
Known methods	One-dimensional LSTM without signal preprocessing	53.8
	A standard example offered in MatLab2020b environment	70.2
	[33]	79.0
	[34]	83.0
	[35]	82.0
Proposed methods	**First simulation**	**87.5**
	Second simulation	87.4

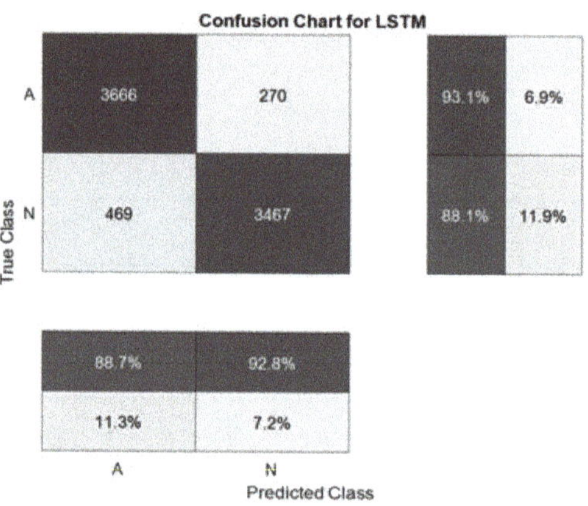

Figure 10. A confusion matrix is a result of training a system for neural network determination of atrial fibrillation on ECG signals with wavelet-based preprocessing.

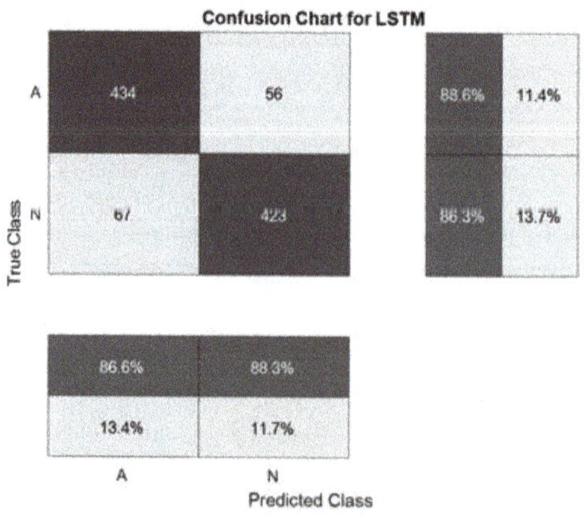

Figure 11. A confusion matrix is a result of the validation of a system for neural network determination of atrial fibrillation on ECG signals with wavelet-based preprocessing.

5. Discussion

The article presents a system for neural network determination of atrial fibrillation on ECG signals with wavelet-based preprocessing. Simulation of the proposed system made it possible to achieve a recognition accuracy of 87.5%, which is significantly higher than the 79.0–82.0% level that can be achieved using known systems [33–35].

Work [33] is devoted to the development of a smartphone application for the detection of atrial fibrillation. The method proposed in [33] uses preliminary preprocessing of the ECG signal, which includes the analysis of the R-R and P-peak intervals. Despite the identity of the methods used, the authors of the work indicate the accuracy of the system at 79.0%, which is significantly lower than the result obtained by modeling the system proposed in this work.

The authors of [34] used the training of a recurrent neural network without preliminary signal processing and achieved an accuracy of 82.0%. The recurrent network is the basis of the LSTM network used in our method, which makes it possible to make comparisons.

In [35], the authors used a part of the MIT-BH 2017 database consisting of normal ECG signals, signals with fibrillation, and "others." A group of noisy signals was not used. For the experiment, the authors used signals with a length of 4 heart counts (4 R-R intervals). The method proposed in [35] did not give a positive result for signals of the "other" category. The results shown for determining the presence of fibrillation from groups with normal signals and signals with fibrillations were 82.0% accuracy, which is also lower than the result of the accuracy of the proposed system for neural network determination of atrial fibrillation on ECG signals.

There are other ways to measure atrial fibrillation. In [36], a more complex methodology is used that requires more computing power. Using multiple convolutional neural networks and LSTM networks is a resource-intensive method. At the same time, the use of pre-processing of the signal before training the neural network can reduce resource costs.

The proposed stages of preliminary processing of ECG signals made it possible to prepare data for further analysis to conduct an automated determination of atrial fibrillation. The average accuracy of a cardiologist's diagnosis by visual analysis of ECG signals is 65.0–70.0% [37]. The use of the proposed neural network system for determining atrial fibrillation from ECG signals by specialists will make it possible to increase the efficiency of diagnostics in comparison with methods of visual diagnosis. The proposed system for neural network determination of atrial fibrillation on ECG signals can only be used as an additional diagnostic tool by specialists. This system is not a medical device and cannot independently diagnose patients.

A promising direction for further research is the construction of more complex systems for the neural network classification of ECG signals, using, together with the analysis of signals, various metadata about patients, such as age, gender, race, genetic predisposition, and other descriptors. One of the next steps for further research is the creation of a mobile application for real-time fibrillation detection. Additionally, further research plans include the complication of the proposed system by using convolutional neural networks to improve the accuracy of determining atrial fibrillation.

Author Contributions: Conceptualization, P.L.; methodology, P.L.; software, M.K.; validation, M.K.; formal analysis, P.L.; investigation, M.K.; resources, U.L.; data curation, P.L.; writing—original draft preparation, M.K.; writing—review and editing, U.L.; visualization, M.K.; supervision, U.L.; project administration, P.L.; funding acquisition, P.L. All authors have read and agreed to the published version of the manuscript.

Funding: This research was funded by the Russian Foundation for Basic Research (project no. 19-07-00130 A) and by the Presidential Council for grants (project no. MK-3918.2021.1.6).

Institutional Review Board Statement: Not applicable.

Informed Consent Statement: Not applicable.

Acknowledgments: The authors are grateful to the North Caucasus Federal University for supporting the competition of scientific groups and individual scientists of the North Caucasus Federal University.

Conflicts of Interest: The authors declare no conflict of interest.

References

1. Wang, L.H.; Chen, T.Y.; Lee, S.Y.; Yang, T.H.; Huang, S.Y.; Wu, J.H.; Fang, Q. A wireless ECG acquisition SoC for body sen-sor network. In Proceedings of the 2012 IEEE Biomedical Circuits and Systems Conference (BioCAS), Taiwan, China, 28–30 November 2012; pp. 156–159.
2. Buxi, D.; Berset, T.; Hijdra, M.; Tutelaers, M.; Geng, D.; Hulzink, J.; Van Noorloos, M.; Romero, I.; Torfs, T.; Van Helleputte, N. Wireless 3-lead ECG system with on-board digital signal processing for ambulatory monitoring. In Proceedings of the 2012 IEEE Biomedical Circuits and Systems Conference (BioCAS), Taiwan, China, 28–30 November 2012; pp. 308–311. [CrossRef]
3. Martinez, J.P.; Almeida, R.; Olmos, S.; Rocha, A.P.; Laguna, P. A wavelet-based ECG delineator: Evaluation on standard da-tabases. *IEEE Trans. Biomed. Eng.* **2004**, *51*, 570–581. [CrossRef]
4. Halhuber, M.J.; Günther, R.; Ciresa, M. Technique of ECG Recording. In *ECG-An Introductory Course A Practical Introduction to Clinical Electrocardiography*; Springer: Berlin/Heidelberg, Germany, 1979; pp. 141–145.
5. Berbari, E.J.; Lander, P. The methods of recording and analysis of the signal averaged ECG. In *Signal Averaged Electrocardiography*; Springer: Dordrecht, The Netherland, 1993; pp. 49–68.
6. Thirrunavukkarasu, R.R.; Meeradevi, T.; Ravi, A.; Ganesan, D.; Vadivel, G.P. Detection R Peak in Electrocardiogram Signal Using Daubechies Wavelet Transform and Shannon's Energy Envelope. In Proceedings of the 2019 5th International Conference on Advanced Computing & Communication Systems (ICACCS), Coimbatore, India, 15–16 March 2019; pp. 1044–1048. [CrossRef]
7. Lippi, G.; Sanchis-Gomar, F.; Cervellin, G. Global epidemiology of atrial fibrillation: An increasing epidemic and public health challenge. *Int. J. Stroke* **2020**, *16*, 217–221. [CrossRef] [PubMed]
8. Goldberger, A.L.; Goldberger, Z.D.; Shivikin, A. *Clinical Electrocardiography: A Simplified Approach E-Book*; Elsevier Health Sciences: London, UK, 2017.
9. Ahuja, A.S. The impact of artificial intelligence in medicine on the future role of the physician. *PeerJ* **2019**, *7*, e7702. [CrossRef]
10. Huang, M.L.; Wu, Y.S. Classification of atrial fibrillation and normal sinus rhythm based on convolutional neural network. *Biomed. Eng. Lett.* **2020**, *10*, 183–193. [CrossRef]
11. Hasan, N.I.; Bhattacharjee, A. Deep Learning Approach to Cardiovascular Disease Classification Employing Modified ECG Signal from Empirical Mode Decomposition. *Biomed. Signal Process. Control.* **2019**, *52*, 128–140. [CrossRef]
12. Yildirim, Ö. A novel wavelet sequence based on deep bidirectional LSTM network model for ECG signal classification. *Comput. Biol. Med.* **2018**, *96*, 189–202. [CrossRef] [PubMed]
13. Faust, O.; Shenfield, A.; Kareem, M.; San, T.R.; Fujita, H.; Acharya, U.R. Automated detection of atrial fibrillation using long short-term memory network with RR interval signals. *Comput. Biol. Med.* **2018**, *102*, 327–335. [CrossRef]
14. Mundhe, P.; Pathrikar, A.K. An overview of implementation of efficient QRS Complex detector with FPGA. *Int. J. Adv. Res. Comput. Commun. Eng.* **2013**, *2*, 4041–4043.
15. Pan, J.; Tompkins, W.J. A real-time QRS detection algorithm. *IEEE Trans. Biomed. Eng.* **1985**, *3*, 230–236. [CrossRef] [PubMed]
16. Zhang, F.; Lian, Y. QRS Detection Based on Multiscale Mathematical Morphology for Wearable ECG Devices in Body Area Networks. *IEEE Trans. Biomed. Circuits Syst.* **2009**, *3*, 220–228. [CrossRef]
17. Abo-Zahhad, M.; Ahmed, S.M.; Zakaria, A. An Efficient Technique for Compressing ECG Signals Using QRS Detection, Estimation, and 2D DWT Coefficients Thresholding. *Model. Simul. Eng.* **2012**, *2012*, 1–10. [CrossRef]
18. Anant, K.S.; Dowla, F.U.; Rodrigue, G.H. Detection of the electrocardiogram P-wave using wavelet analysis. *Int. Soc. Opt. Photonics* **1994**, *2242*, 744–750. [CrossRef]
19. Yang, H.; Bukkapatnam, S.; Komanduri, R. Nonlinear adaptive wavelet analysis of electrocardiogram signals. *Phys. Rev. E* **2007**, *76*, 026214. [CrossRef] [PubMed]
20. Kumar, A.; Komaragiri, R.; Kumar, M. Design of wavelet transform based electrocardiogram monitoring system. *ISA Trans.* **2018**, *80*, 381–398. [CrossRef]
21. Fujita, H.; Cimr, D. Computer Aided detection for fibrillations and flutters using deep convolutional neural network. *Inf. Sci.* **2019**, *486*, 231–239. [CrossRef]
22. Bnou, K.; Raghay, S.; Hakim, A. A wavelet denoising approach based on unsupervised learning model. *EURASIP J. Adv. Signal. Process.* **2020**, *2020*, 1–26. [CrossRef]
23. Chavan, M.S.; Mastorakis, N.; Chavan, M.N.; Gaikwad, M.S. Implementation of SYMLET wavelets to removal of Gaussian additive noise from speech signal. In Proceedings of the Recent Researches in Communications, Automation, Signal Processing, Nano-technology, Astronomy and Nuclear Physics: 10th WSEAS International Conference on Electronics, Hardware, Wireless and Optical Communications (EHAC'11), Cambridge, UK, 20–22 February 2011; p. 37.
24. Luo, G.; Zhang, D. Wavelet Denoising. *Adv. Wavelet Theory Appl. Eng. Phys. Technol.* **2012**, 634. [CrossRef]
25. Zarei, A.; Asl, B.M. Automatic Detection of Obstructive Sleep Apnea Using Wavelet Transform and Entropy-Based Features from Single-Lead ECG Signal. *IEEE J. Biomed. Health Informatics* **2018**, *23*, 1011–1021. [CrossRef]
26. Donoho, D.L.; Johnstone, J.M. Ideal adaption by wavelet shrinkage. *Biometrika* **1994**, *81*, 425–455. [CrossRef]

27. Boashash, B. Estimating and interpreting the instantaneous frequency of a signal. I. Fundamentals. *Proc. IEEE* **1992**, *80*, 520–538. [CrossRef]
28. Carson, J.R.; Fry, T.C. Variable Frequency Electric Circuit Theory with Application to the Theory of Frequency-Modulation. *Bell Syst. Tech. J.* **1937**, *16*, 513–540. [CrossRef]
29. Van der Pol, B. The fundamental principles of frequency modulation. *J. Inst. Electr. Eng. Part III Radio Commun. Eng.* **1946**, *93*, 153–158. [CrossRef]
30. Hochreiter, S.; Schmidhuber, J. Long Short-Term Memory. *Neural Comput.* **1997**, *9*, 1735–1780. [CrossRef]
31. PhysioNet Computing in Cardiology Challenge 2017 (CinC Challenge). Available online: http://physionet.org/challenge/2017/ (accessed on 13 March 2021).
32. Mahmoodabadi, S.Z.; Ahmadian, A.; Abolhasani, M.D.; Eslami, M.; Bidgoli, J.H. ECG Feature Extraction Based on Multi-resolution Wavelet Transform. In Proceedings of the 2005 IEEE Engineering in Medicine and Biology 27th Annual Conference, Shanghai, China, 1–4 September 2005; pp. 3902–3905.
33. Andreotti, F.; Carr, O.; Pimentel, M.A.F.; Mahdi, A.; De Vos, M. Comparing Feature Based Classifiers and Convolutional Neural Networks to Detect Arrhythmia from Short Segments of ECG. In Proceedings of the 2017 Computing in Cardiology (CinC), Rennes, France, 24–27 September 2017. [CrossRef]
34. Billeci, L.; Costi, M.; Lombardi, D.; Chiarugi, F.; Varanini, M. Automatic Detection of Atrial Fibrillation and Other Ar-rhythmias in ECG Recordings Acquired by a Smartphone Device. *Electronics* **2018**, *7*, 199. [CrossRef]
35. Wang, J.; Li, W. Atrial Fibrillation Detection and ECG Classification based on CNN-BiLSTM. *arXiv* **2020**, arXiv:2011.06187.
36. Jin, Y.; Qin, C.; Huang, Y.; Zhao, W.; Liu, C. Multi-domain modeling of atrial fibrillation detection with twin attentional convolutional long short-term memory neural networks. *Knowl. Based Syst.* **2020**, *193*, 105460. [CrossRef]
37. Zheng, Z.; Chen, Z.; Hu, F.; Zhu, J.; Tang, Q.; Liang, Y. An automatic diagnosis of arrhythmias using a combination of CNN and LSTM technology. *Electronics* **2020**, *9*, 121. [CrossRef]

MDPI
St. Alban-Anlage 66
4052 Basel
Switzerland
Tel. +41 61 683 77 34
Fax +41 61 302 89 18
www.mdpi.com

Applied Sciences Editorial Office
E-mail: applsci@mdpi.com
www.mdpi.com/journal/applsci

www.ingramcontent.com/pod-product-compliance
Lightning Source LLC
LaVergne TN
LVHW070617170726
843515LV00004B/113

* 9 7 8 3 0 3 6 5 5 5 1 5 7 *